Vinícius Arcanjo Monteiro

Geography Postcards

Vinícius Arcanjo Monteiro

Geography Postcards

Fieldwork in Geography

ScienciaScripts

Cover image: www.ingimage.com

This book is a translation from the original published under ISBN 978-620-2-19326-9.

Publisher:
Sciencia Scripts
is a trademark of
Dodo Books Indian Ocean Ltd. and OmniScriptum S.R.L publishing group

120 High Road, East Finchley, London, N2 9ED, United Kingdom
Str. Armeneasca 28/1, office 1, Chisinau MD-2012, Republic of Moldova, Europe
Printed at: see last page
ISBN: 978-620-7-27591-5

SUMMARY

The points represent our destinations, dreams and arrivals. The lines, the paths we have to choose to reach our destinations, dreams or points of arrival. But the paths are also destinations and must be observed along the way. Between one point and another, there are the blurs. These are the spaces for possibilities. In The Little Prince, arriving at his sixth country, I explain the poetics of paths and the poetic moments of journeys. Geographers map rivers, lakes, mountains and depressions. But if the geographer isn't an explorer and lets himself be carried away by the paths, he won't be able to define the lines that find the next point.

Expedition. The path explored reveals other paths to follow. Better, clearer paths. This way, we can take our friends and the bag of knowledge from the whole exploration with us. The point of arrival may seem less desirable than at the start of the expedition. From that point, there are new points and new lines to follow to reach the point. But it's not that the point is worse than I thought, it's that the expedition, with its attentive eye on everyone and everything around us, brings knowledge, experiences and people who let us "experience".

In the web of lines, the people who let us go and make us who we are. Each person is given experiences that touch us, pass through us and change us. It's important to be clear about the people who enhance our experiences and help us grow by keeping them with us. The others, let them be valid experiences so that we don't repeat mistakes.

Reflecting on everything is something I've added to my life. Take a step back every day, including from yourself, and reflect on all the points you want to. As for the lines, be attentive and watch the path. It's an expedition!

By Vinicius Arcanjo Monteiro.

SUMMARY

Inspired by studies in Cultural Geography, this paper presents the results of a study interested in outlining the importance of Fieldwork in Geography for the training of geographers and the circumstances surrounding this methodology in the contemporary period. For the purposes of this study, we identified the main data from the Boletim Paulista de Geografia (No. 84), which led us to reflect on the similarities in the discussions that serve as theoretical and methodological subsidies for Fieldwork in/for Geography. The themes and concepts collected indicate that, despite the plurality of experiences in Fieldwork, it remains in constant construction, evolving and constituting itself according to the construction of geographic science, thus allowing us to advance in the analysis of Fieldwork in Cultural Geography, that is, the systematization of humanistic ideas and their applicability in Fieldwork. To this end, the "Geography Postcards" project will be proposed as a way of showcasing fieldwork experiences in geography.

Keywords: Fieldwork; Humanistic-Cultural Geography; History of Geographical Thought; Expedition/Exhibition.

INTRODUCTION

During the geographer's training, fieldwork is present as a legacy of experiences and findings that show, above all, part of his work and usefulness in the construction of his science. Nowadays, the transformation of space has become more rapid and increasingly articulated (SERPA, 2006. p. 10), requiring those who wish to study it to become aware of their role and that of those who, directly or indirectly, participate in this process, and who seek to understand the natural, social, economic and cultural dynamics of space.

Through Humanistic Geography, it is understood that the method discusses geographical phenomena with the aim of achieving a better understanding of man and his condition, thus seeking "an understanding of the human world through studies of social relations with nature, their geographical behavior, as well as their feelings and ideas about space and place" (TUAN, 1976):

Scientific approaches to the study of man tend to minimize the role of human awareness and knowledge. Humanistic Geography, in contrast, specifically tries to understand how geographical activities and phenomena reveal the quality of human awareness. (TUAN, 1976 p. 143)[1]

From this perspective, this research aims to discuss the importance of fieldwork. Based on the classic publication of the Boletim Paulista de Geografia (No. 84, 2006)[2] which, in this special issue, deals with the theme of Fieldwork. The journal contains 7 articles on the role of the geographer in fieldwork, the use of concepts and theories combined with practice, among other subjects.

It is worth mentioning the Brazilian articles analyzed such as: Ângelo Serpa, Ricardo Baitz, Paulo Alentejano and Otàvio Miguez da Rocha Leao, Luis Antônio Bittar Venturini and Valéria de Marcos, and French geographers such as Bernard Kaiser and Yves Lacoste, who pointed out and proved to be primary sources in the analysis of Fieldwork. In accordance with Alentejano, it is of the utmost importance to support:

The contribution of the Association of Brazilian Geographers (AGB) to the dissemination of Field Work as a fundamental tool for geographers, since its first congresses represented a privileged space for the production of Brazilian Geography, with emphasis on the excursions it promoted, the lectures it organized and the texts it published in the Boletim Geogràfico. (ALENTEJANO, 2002, p. 146 apud ALENTEJANO, 2006. p. 54)

In addition to this Bulletin, we have consulted some articles that explore the theme of Fieldwork, but as it is a subject that does not have many reference works, it is necessary to study this topic. Firstly, a retrospective of fieldwork in geography will be carried out, starting with the four schools

1 Yi Fu Tuan - Chapter transcribed from Annals of the Association Of American Geographers, 66: (2), July 1976. Original title: Humanistic Geography. Translated by Maria de Queiròz. TUAN, Y. F. Humanistic Geography. In: CHRISTOFOLETTI, A. (Org.). Perspectivas da Geografia. Sao Paulo: Difel, 1982. Chap. 7, p. 143-164

2 Boletim Paulista de Geografia (No. 84; 2006), which began its activities in 1949 and is still in circulation today, with a total of **91** editions and which, in this special edition, deals with the theme of Field Work. (ALVES, 2010 p. 34)

of geographical thought (classical-traditional geography, quantitative-theoretical geography, critical geography and humanistic geography), highlighting the main methods, techniques and concepts used at each time, based on the understanding that:

> Research into the history of geographic thought is of great importance for evaluating the behavior of the science; verifying the evolution and trends of studies guarantees changes in theories, concepts and methodologies in geography. Geographical science has undergone, and is still undergoing, major transformations in terms of the use of methods, theories and concepts, and the study of these changes is necessary if we are to understand the direction in which geography is heading. (ALVES, 2010)

Subsequently, interviews were conducted with teachers from the Geography course at the Institute of Natural Sciences - UNIFAL-MG, covering issues pertinent to the topic. With these analyses, we built correlations between theory and academic practice, as well as debating the role of Field Work in teaching Geography at school level, exploring points of articulation between university and society and theoretical deepening, since it is in the university's interest to train for research:

> If we don't want geography to disappear from universities because of its students, we need to make their training much more research-oriented, right from the start of their studies. Those who can still find a place in *high schools and* colleges will only be the best teachers and will be able to initiate their students into practical research, which is politically very useful for all future citizens. If the research relationship is to change little by little, the vast majority of young people must be initiated into research and fieldwork at university, so that they can (...) explain its usefulness to people, why it is necessary to know the results and how they can be used. (LACOSTE, 2006, p. 86)[3]

Based on the finding that teachers credit the involvement/development of knowledge through fieldwork and that fieldwork generally achieves the results expected by the subject, I understand that we need to go further, proposing solutions that at least indicate the systematization of humanistic ideas and their applicability in fieldwork, supported by the concept of *exposing* experiences in geography fieldwork.

A better understanding of the concept of **expedition/exposure**, which until then had also been worked on by the authors of the Boletim Paulista de Geografia, is best organized in the words of Yves Lacoste, based on the ramification of the results of Field Work:

> The expedition/exhibition seems to be a very effective method of training students for research. (...) The students who have taken part in these expeditions/exhibitions have mostly experienced a certain enthusiasm (especially if they are activists) at the idea of reporting the results of their research to the public. (LACOSTE, 2006, p. 90)

It is important to emphasize that from an initiation to research, the Fieldwork presented, based on the interaction of the media in the university-society or university-school relationship, suggests the

---

3 Yves Lacoste, In: Boletim Paulista de Geografia (2006): "Research and fieldwork: a political problem for researchers, students and citizens", is a text originally published in French in 1977 by the magazine Heròdote (n° 8, Oct./Dec. pages 3 to 20). Original title: "L'enquête et le terrain: un ploblème politique par le churchus, les éstudiants at les cittoyens".

expansion of the educational space, which, through this, could have repercussions on other levels of scale in a more liquid and universal way, and be the beginning of a real expedition.

- Objectives

- **General Objective**

Discuss the importance of fieldwork in the production of geographical knowledge.

- **Specific Objectives**

S Content analysis of the theory and concepts of fieldwork in the evolution of geographical thought;

S Understanding Fieldwork in teaching practice and in the training of the geographer;

S Analyze the procedures for systematizing field research in geography, with an emphasis on cultural geography.

- Methodological Procedure

The research into the importance of fieldwork in geography is based on the classic publication Boletim Paulista de Geografia (No. 84, 2006) which, in this special issue, deals with the theme of fieldwork. The journal contains 7 articles on the role of the geographer in fieldwork, the use of concepts and theories combined with practice, among other subjects.

In addition to this Bulletin, we have consulted some articles that explore the theme of Fieldwork, but as it is a subject that does not have many reference works, it is necessary for us to study this theme. Firstly, we'll take a look at fieldwork in geography, starting with the four schools of geographic thought (classical-traditional geography, quantitative-theoretical geography, critical geography and humanistic geography), highlighting the main methods, techniques and concepts used at each time.

This will serve, among other things, to correlate technical advances with the types of methods used in fieldwork; to relate the political and historical context to the techniques and methods used by geographers at any given time. It will also help these schools to define the importance of fieldwork in the construction of their results.

Subsequently, interviews were conducted with five teachers from the Geography course at the Institute of Natural Sciences - UNIFAL-MG, covering issues pertinent to the topic. The aim of these analyses is to build correlations between theory and academic practice, as well as to debate the role of Fieldwork in teaching Geography at school level.

Finally, there will be a conclusive analysis of the importance of Fieldwork and Cultural Geography, from a Humanist perspective, i.e. the systematization of humanistic ideas and their applicability in Fieldwork.

DEVELOPMENT

CHAPTER 1

The Evolution of Fieldwork in Geographical Thought

The first chapter is interested in constructing the moments of geographical knowledge through the theme. This is because the science of geography arose in the midst of a collection of travelers' reports and expeditions aimed at describing the earth's surface, and these works gave rise to the importance of fieldwork. We will analyze the methods used in fieldwork from the beginning of geographic thought to the humanistic/cultural geographic moment.

The expeditions of the naturalists are the "roots" of Field Work, in other words, concomitant with the birth of geographic science, which means that the observation of the landscape, as an aesthetic manifestation of reality, was already actively involved in the formation of geographic thought, alongside reports and descriptions. "In images, both the point of view and the space seen are relative and subjective. More than either cosmography or geography, chorography makes room for the imaginative and creative role of the individual," wrote Denis Cosgrove (1999, p. 107 apud MARQUES, 2009, p. 71). Thus, when "[...] words are lacking to talk about the forms of relief or rocks", geographers resort to the expressive techniques of diagramming, engraving and painting, mixing scientific knowledge and aesthetic emotion in engravings highlighting physical factors (figure 1): (MARQUES, 2009):

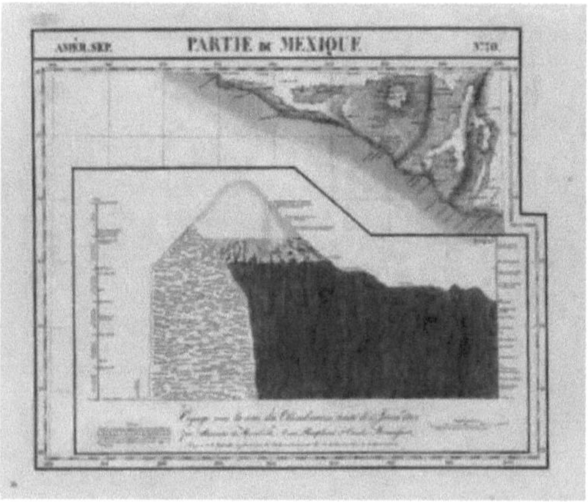

Figure 1: Alexander von Humboldt. *Plan, topography and altitude map* of *Mount Chimborazo.* 1807. Source: http://www.humbot.org, (apud MARQUES, 2009 p. 73)

With the publication of the first volume of *Anthropogeography* by the German Friedrich Ratzel, in 1882, an attempt was made to delineate the relationship between geographical science and the men who write these graphs. Ratzel's determinism was followed by the possibilist thinking of Frenchman Paul Vidal de la Blache at the turn of the 20th century. Responsible for the legacy of thought on the *unity of the earth,* Paul Vidal de la Blache wrote a work in which geography was made up of natural history, the physics of the earth. This thinking underpins the development of modern geography. The so-called post-modern geographies, on a smaller scale than the still predominant modern geographies, investigate the relationships between the historical production of space and its political, economic and cultural implications, drawing closer to anthropology, sociology, philosophy and the arts (MARQUES, 2009 p. 25).

To understand how Fieldwork behaves through method, we recall Eliseu Sposito when he explains that a method means: "a set of rational procedures, based on rules, aimed at achieving a specific objective. So a scientific method presents itself as laws and categories, which are historically related to specific procedures and theories disseminated by the scientific community" (SPOSITO, 2004 p. 25).

By understanding a method in this way, it can be seen that during this period, the methods of analysis in geography were born to describe the Earth and to point out its diversity. As we move forward, geographers are now trying to answer questions that were considered groundbreaking at the time: To what extent is the destiny of peoples linked to the country in which they live? Does it influence people? Or is there a subtle harmony between the natural order and the social order?

The relationship between society and the environment became central to the discipline. Anthropogeography, named by Frederich Ratzel and adopted by the majority of young researchers. In France, Vidal de La Blache and the group he stimulated, translated the term human geography, which was derived from analysis of human evolution. The term Human Geography and the new notion took hold in the last years of the 19th century. (CLAVAL, 2007, p. 7)

So, when scientific geography emerged, the main techniques used in fieldwork were observation, sketching, comparison, description, in other words, simple techniques, without a large technological apparatus, which was supplied by the time spent involved in the place under analysis, what La Blache calls Regional Monography.

From a Lablachian perspective, a regional monograph should contain a detailed analysis of the physical environment, the forms of occupation, human activities and how man adjusts to nature. The look at nature should include a historical perspective in the analysis of the relationship between man and the environment. Fundamentally, the regional monograph should establish the integration of physical and social elements and add a synthetic view of the region. (LENCIONI, 2003, p.105 Apud ALVES, 2010, p. 37).

The Traditional Geographer's routine was also "to observe, locate and describe", highlighting Fieldwork as an important area of geography as a science. "The wealth of detail in the countless regional monographs predominated in the 1940s and 1960s, becoming a reference and a key stage in geographic methodology." (ALVES, 2010. p. 39)

Geographers at this time began to develop a science based on regional differentiations on the earth's surface, based on the analysis of research and field reports drawn up by travelers and naturalists concerned with conquering territories. This fact explores the beginnings of Geography, pointing to

Field Work and the descriptive method as being largely fundamental to the working method of Geographers and the constitution of the science (ALENTEJANO, 2006, p. 53).

Thus strengthened by naturalist views, Fieldwork accompanied the transformations of the capitalist system, affecting Geography right up to the present day, and yet it is still the main common point among geographers. According to Venturi (2006):

> [...] the field is where the complexity of reality is revealed and brought to the understanding of the geographer, armed with his main concepts, such as landscape, space, region and place, for example, which materialize in reality, give it meaning and derive meaning from it. This is where the academic boundaries of the disciplines cease to make sense and are replaced by countless connections between the facts observed, in a process of conceptual reconstruction.

In Geography's descriptive method, the object took precedence over the subject, i.e. the object studied was the "target", observed from above, influencing the researcher and their knowledge. Although there was a tendency towards the "neutrality of science" (an assumption), reality was described by means of hypotheses and deduction.

The paradigm shift from Traditional Geography to Theoretical / Quantitative Geography was due to the doubts and lack of precision of some descriptive works, which were based only on the "observed object". (ALVES, 2010 p. 139). Affirming the revolution, it was based on geography's need for a theoretical-methodological framework to support it as a science. In this sense [...] quantitative methods in geography represent a new and powerful weapon for the analysis of geographical phenomena, capable of making geography a branch of human knowledge equal to others of a scientific nature, due to its ability to specify phenomena and establish general principles according to which they occur. (GALVÂO & FAISSOL, 1970, p.5 apud ALVES, 2010. P. 139) and which, according to Paulo Alentejano:

> All this accumulation was thrown into disarray from the 1960s onwards, when, in the wake of the hegemony of Theoretical and Quantitative Geography, Field Work was execrated and practically wiped off the map of geographers' practices, on the grounds that information technology and mathematical models would be more appropriate tools for investigating reality." (ALENTEJANO, 2006, p. 55)

Claval takes an interesting approach by revealing that "geography has occupied an important place in cultural realities since its birth, but has captured them through a reductionist lens", according to CLAVAL (2007): due to the emphasis placed on techniques, tools and the transformation of landscapes.

Brazilian Geography, on the other hand, reveals that the second moment in the history of geography, when fieldwork ceased to be paramount, that is, its analysis centered on the interpretation of secondary data, above all the use of the Censuses of the Brazilian Institute of Geography and Statistics (IBGE). Among the techniques employed are the use of statistics, thematic maps, spatial-localization models, among others.

> Fieldwork is the collection of statistical data from secondary sources, based on variables compiled by censuses or other institutional surveys, although these surveys present a quantifiable unit, (...) the basis from which situations are compared using quantitative techniques. (ROCHA, 1999) With data from institutional censuses, the researcher does not really have the power to choose the variables analyzed, (...) often the data is collected using techniques and procedures that prove to be incompatible with the research questions. (apud SERPA, 2006. p. 14)

However, this is not to say that the studies of this moment did not advance in the very constitution of "geographical identity", but it does indicate a gap in human conception, since many of the works were formal formations of analysis. Many of the contributions are "momentary" technical advances that are surpassed by new perspectives.

Thus, from the 1970s onwards, the research model in geography, based on statistics, was severely questioned because it left out social, political and cultural issues. Historicity gained ground in fieldwork. The historical-dialectical approach represents the aim of understanding and unveiling social reality. This approach allows history to be rescued and the contradictory movement of ideas and the actors and agents involved to be seen. According to Alves (2010) on the structure of Marxist historical materialist thinking:

> The basis of the materialist-historical system is given by the rules that determine the type of production relationship in relation to the development of the productive forces; these two elements are the fundamental terms that define a mode of production and, at the same time, constitute the cause of dialectical transformation (GOMES, 2005, p.283). The dialectical method has various nuances and distinctions, from the dialectics of Hegel's idealism to the historical-dialectical materialism of Marx and Engels. However, the Marxist perspective has been hegemonic in this thinking in Brazil (apud ALVES, 2010. P. 39/40).

Thus, in the dialectical method, the subject is constructed and transformed in relation to the object and vice versa. The work that uses this method is characterized by being more critical of reality, by its concreteness and by the fact that it shows the historical contradiction that exists in the object studied (SPOSITO, 2004).

Just as the initial movement of Critical Geography, under the radical critique of the empiricism dominant in Traditional Geography, led to a denial of the validity of Fieldwork as an instrument for constructing geographic thought, due to the emphasis placed on Marxist theory (ALENTEJANO, 2006). The difficult alliance between theory and instrumental practice has to be pursued and always grounded in field research. (RUA, 1997, p. 45 apud ALENTEJANO, 2006. p. 55)

Ângelo Serpa contributes to demystifying this process which, according to the author, if we think from a dialectical perspective, we are in the field of structuring information and aggregated data, with a greater *tendency towards homogeneity,* while through phenomenology, we enter the field of factual information and individual data, with a greater *tendency towards heterogeneity*. (SERPA,

2006. p. 18)

It consists of admitting that there is no privileged method when we adopt a unit of observation and assign attributes that are associated with geographical knowledge, because according to Lacoste, "None of them is sufficient when we consider some space for observation, we apprehend such phenomena in the same proportion that it deforms or conceals others and other structures, of which one cannot prejudge the role and therefore cannot neglect it."

Thus, at the beginning of the 1980s, the humanistic perspective was introduced into Brazilian geography, converging field research towards an analysis of place which, in turn, used phenomenology to value the lived experience of the subject, describing the quality of the emotion experienced in specific cases. In the phenomenological method, the subject describes the object and its relationships from their point of view, after taking intellectual ownership (SPOSITO, 2004). Important techniques include oral history and participatory research:

> The scientist and his science must first be a moment of commitment and participation in the historical work and projects of struggle of the other, whom, rather than knowing in order to explain, the research aims to understand in order to serve. From then on, a new coherence of scientific work is established and allows different techniques to be used in the service of the method that constitutes it: the report of other observers, even non-scientists, the reading of documents, the application of questionnaires (...), the observation of life and work. Participatory research was invented (BRANDÂO 1987, p. 12 Apud MARCOS, 2006. p. 108-109).

According to Yi-Fu Tuan, the humanistic method represents the scientific researcher (and science) because it critically and reflexively builds on scientific knowledge. (...) "The humanistic geographer must be acutely aware of the obstacles to human freedom; concerned with the quality of human awareness and learning" (TUAN, 1976) and collaborates on questions pertinent to a science:

> How do people acquire spatial skills and knowledge? How do people become emotionally involved with a place? These questions show that the humanist geographer shares the concern of the behavioral psychologist. Their questions are similar, although they are addressed to phenomena of different complexity and scale. Even their field techniques have in common the detailed observation of individual behavior in real life. (TUAN, 1976. P 159).

In Anne Buttimer's phenomenology, in which she suggests *grasping the dynamism of the lived world*, and how the representation(s) of human experience are formed in space, she understands the function of "phenomenologists". In the essays on "lived and existential space", they start from the notion that each person has their own natural place, but is surrounded by concentric "layers" of lived space:

> Phenomenology invites us to explore some of the unifying forces in the human experience of the world. Assuming (...) that such conditions may reside in the facets routinely encountered in daily life, this notion offers a good dialog between phenomenology and geography. (...) "World" for the phenomenologist is the context within which consciousness is revealed. (BUTTIMER, 1976. P. 172)

Based on these humanistic and phenomenological reflections, fieldwork makes sense because it is a process of identification mediated by cultural and social signs, insofar as it brings the scientific value of field research to the place, and in humanistics it is an absolute method of exploration.

1.2 Chapter considerations:

The reality and essence of spaces can be seen through the communication of methods, so Ângelo Serpa discusses the possibility of both phenomenology and dialectics working together in fieldwork as complementary strategies, always seeking to build a synthesis between subject and object:

After all, there should be no incompatibility, in terms of procedure, between an analysis of the individuality of phenomena and the recovery of their historical dimension, nor even between the conception of science interested in history and in the mediation of man-nature and another focused on the essence of phenomena" (SERPA, BPG, 2006. P 29).

Graph 1 attempts to synthesize the "variables" "advances" and "setbacks" in the use of fieldwork at different times in geography, suggesting the evolution (maturing) of methods in relation to the usefulness of fieldwork. The naturalist expeditions are the "roots" of the science that has been advancing its conception based on the relationship between man and the environment and their complete interaction over the centuries (XVIII - XXI), as suggested by the analysis of graph 1.

However, the progressive advance of Fieldwork does not indicate the superiority of one method over the other; field analysis is part of the methods. Coming from a traditional method, we can still see the topicality of the discussion, pertinent in the construction and in absorbing the responsibilities of this construction in Fieldwork given the observation, apprehension and criticism of the object it provides.

According to the sources, fieldwork is characterized as an experimental tool at the service of geographers and also as a traditional way of doing geography. Thus, after natural confessions, we can think that Fieldwork, offering this contact with reality and being a characteristic of geography, whether for carrying out scientific research, (using techniques, technologies and instruments or just immediate observation) individually or collectively, is an essential project for the geographer's training and for the confession of reality in contemporary geographical knowledge.

Graph 1:

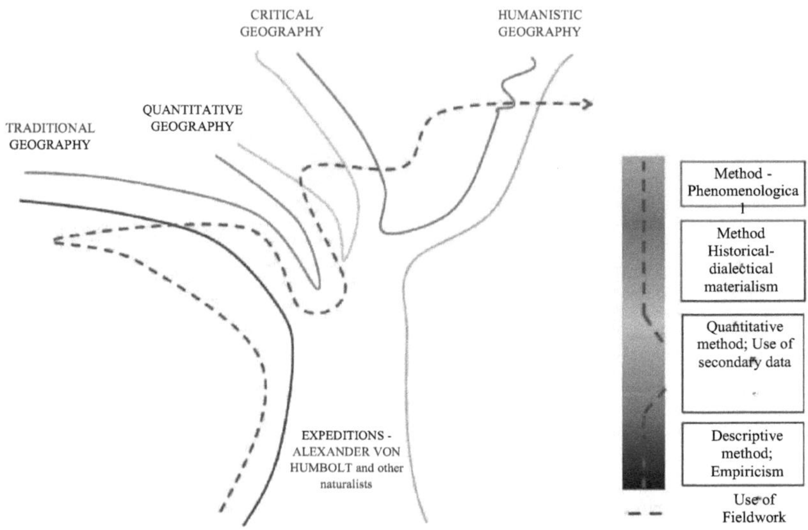

THE EVOLUTION OF FIELDWORK

IN THE CONSTRUCTION OF GEOGRAPHICAL THOUGHT

Organization: Vinicius Arcanjo Monteiro, 2014.

CHAPTER 2

Boletim Paulista de Geografia issue 84: Theoretical and practical reference for fieldwork

The research on the importance of Fieldwork in Geography comes from the classic publication Boletim Paulista de Geografia* (No. 84, 2006), which in this special issue deals with the theme of Fieldwork. The authors mentioned above are also part of the magazine (7 articles), which discuss the role of the geographer in fieldwork, the use of concepts and theories combined with practice, among other subjects.

The summary of the articles available in the Bulletin (*attached*), are precious sources for assimilating this content, and it is plausible and possible to explore the themes contained, provided for analysis, in order to select / organize the contents in view of the importance that the theme arises in the construction of geographic knowledge and in the Teaching of geography.

The contents of the articles by Ângelo Serpa, Ricardo Baitz, Alentejano, Luis Antônio Bittar Venturi, Yves Lacoste, Bernard Kaiser and Valéria de Marcos were grouped according to concepts, themes, debates and analysis of the theoretical framework. We propose an analysis of the grouped data, assuming the result of generating themes for debate.

2.1. Content analysis

This topic is organized according to the sources of the Boletim Paulista de Geografia (No. 84, 2006), which are summaries of the main discussions in each article analyzed. This is a descriptive analysis of the content of the Boletim Paulista de Geografia, highlighting the concepts and emphasizing the importance of fieldwork.

A complete summary table is included in the *appendix. It* organizes the methods, concepts, themes and debates and the references that each author used in order to compact the bibliographical review.

BPG: Concepts and Themes in the Articles of the Boletim Paulista de Geografia (NÙMERO 84. SÂO PAULO-SP, JUL.2006 - Fieldwork)

Synthetic Analysis: Concepts, Themes and Discussions:

1- Ângelo Serpa

Concepts:

Space; Spatial conceptualization; Landscape; (Vertical/ Cultural); elements; totality.

Topics for debate:

S Theoretical-methodological reflection on the importance of fieldwork in the construction of contemporary geographical knowledge;

S Fieldwork as a specific discipline of geography;

S Articulation of concepts, theories and methodological procedures in geography;

S Fieldwork is the possibility of cutting out, analyzing and conceptualizing space, according to the questions, goals and objectives defined by the research subject;

S Fieldwork as an instrument for overcoming dichotomies in geographical science;

S Theory-method.

- Ângelo Serpa:

SERPA, A. S. P. O Trabalho de Campo em Geografia: Uma Abordagem Teórico- Metodológica. Boletim Paulista de Geografia, v. 84, p. 7-24, 2006.

The article is entitled: *Fieldwork in Geography: A theoretical-methodological approach.* This article seeks to make a theoretical-methodological reflection on the importance of fieldwork for the production of geographical knowledge in contemporary times.

The author relies on four of the presuppositions he considers to be the main ones in geography and thus tries to conceptualize and give meaning to fieldwork in geography:

1st assumption: geography's disciplinary specificity;

2nd assumption: the cut-out of the spaces of conceptualization and the centrality of the concept of space;

3rd assumption: fieldwork to overcome the dichotomies and ambiguities of geography;

4th assumption: theory and methods are inseparable in geography fieldwork.

On the subject of fieldwork as a specific discipline of geography, the author points out that "a significant spatial cutout conceptualizes the geographer in his objective for spatial analysis in line with the next assumption". Through fieldwork, there are various possibilities for cutting out, analyzing and conceptualizing space, according to the questions, goals and objectives defined by the research subject.

Ângelo Serpa suggests geography as a science of space, emphasizing the specificity of geography's scientific productions. In order to value the division of space, making it the central issue in the operationalization of fieldwork to conceptualize reality in coherence with the phenomena we wish to study. The possibility of dividing space is assumed in conjunction with the method and in an interdisciplinary way.

Although he observes the relationships of space, the author indicates the concept of scale as a mediator, thus forming the relationship: *researcher - scale - unit of observation and attributes*;

suspending scale as a fundamental problem of fieldwork, since there is a tendency to homogenize important attributes in the analysis.

> We can therefore conclude that the more abstract sections will favor and give visibility to structuring information, aggregated data, latent phenomena and the tendency towards homogeneity and the model, while the more concrete sections will value factual information, individual or disaggregated data, manifest phenomena and the tendency towards heterogeneity. There is a tendency towards homogeneity as the scale progressively decreases, while heterogeneity increases proportionally as the sections become more concrete and the scale increases. (SERPA, 2006 p. 13)

The author also points out the relationship between the researcher and the fieldwork data. In this way, the research variable must meet the research objective, i.e. the security of working with the desired variable, understanding the primary data sources as the researcher's "possession".

Another assumption pointed out by Ângelo Serpa is the ambiguities that geography still "carries" as a mark of its history. In his words: "Fieldwork must be based on the totality of space, without forgetting the specific arrangements that make each place, city, neighborhood or region a particular articulation of physical and human factors in a fragmented but (increasingly) articulated world."

In this sense, we should note the statement that Fieldwork is an "analysis of the particular in the totality". Observed from the impulse that the author suggests to new geographers to "search not only for similarities but also for differentiations between places, regions, landscapes and territories that express the totality of space", thinking of space as something dynamic and changeable.

Landscape is highlighted as an important element in contemporaneity and worked on as the key to overcoming these ambiguities, as the author works with the concept of *vertical landscape* considering the worldviews privileged in geography through spatial analysis and also uses the cultural landscape as a contribution to the geographer's eye, being at the same time the bearer of its own symbols that interchange with foreign symbols and meanings:

> Landscapes and maps can "lie" if we don't admit that it's not just "objective reality" that should hold our attention, but also how this reality speaks to the senses of the subject who observes and researches. We need to recognize the landscape as a collusion, "exploring its intertwined threads and reciprocal exchanges". Seen in this way, the landscape is both a mark and a matrix, since it seems that "societies organize their environments according to the perception they have of them, and, reciprocally, it seems that they perceive themselves according to the organization they give them". (SERPA, 2006, p. 12)

Thus, when we talk about fieldwork, we can see that a visitor's gaze must be combined with a gaze that has maps, aerial photographs and research on (...) this apparatus requires a geographical gaze in relation to space.

One of the most fundamental parts of this article, Ângelo Serpa makes the following assumption: theory and method are inseparable in fieldwork. It consists of admitting that there is no privileged method when we adopt a unit of observation and assign attributes that are associated with geographical knowledge, because according to Lacoste, "None of them is sufficient when we consider some space for observation, we apprehend such phenomena in the same proportion that it deforms or hides others and other structures, of which one cannot prejudge the role and, therefore, cannot neglect it."

The reality and essence of spaces can be recognized through the communication of methods. The author provides a new methodology and tools for geographers when he discusses the possibility of both phenomenology and Marxist dialectics working together in fieldwork as complementary strategies, always seeking to build a synthesis between subject and object.

By way of conclusion, the author reaffirms fieldwork in geography as the basis for research and the production of geographical knowledge. Reaffirming the central position of the concept of space. Bringing concepts, theories and methodological procedures closer together as an *organic and coherent* analysis of the fieldwork/research carried out by geographers.

2 - Ricardo Baitz

Concepts:

Implication; institutions. exhibition.

Topics for debate:

S Implication as a methodology for Fieldwork

S The implied subject x implied object relationship; (subject-object)

S Approaching the object/researcher as an active subject

S The non-neutrality of science

S Implication and pedagogy

S Implication and participation

- Ricardo Baitz:

BAITZ, Ricardo. The implication: a new sediment to explore in Geography? Boletim Paulista de Geografia, v. 84, p. 25-50, 2006.

With the title: Implication - *a new sediment to explore in geography*, Ricardo Baitz seeks to introduce the applications of "implication" in fieldwork. Presenting the implicated researcher, implication without limits/overimplication; implication and pedagogy; implication and focus in research; (im)possibility of implication, the author suggests this as the central concept of the discussions in this article based on sociology and pedagogy. *"Implication means reunion and its presupposition is the division that must be worked through in order to reassemble the split unit, now on another level, in a complex way, with the parts sewn together from within."* (BAITZ, 2006, p. 32)

With the exception of a few works, field research is a presupposition in geography. Transcription (description), representation (cartography) and modification (planning) are all activities carried out by geographers that refer, at some point, to work on the ground.

The appropriation of science for this and technique, initially points to a discussion of the "splitting" of the world into parts, which pointed to "progress" insofar as it left the *primitive stage of separation and moved on to* articulation, which was initially external and later internal, arriving at dialectics.

However, he notes that by leaving the office and going into the field, the scientists (ethnologists) broke with the ethics established until then, which was considered audacious at the time, Baitz reveals. This practice revealed a new objectivity to science, as follows: "Going into the field and feeling it made research more complete and complex, as well as raising questions that had hitherto been obscured". (BAITZ, 2006. p. 29)

The process of knowledge is allied to the junction of *subject and object, in other* words, without hierarchizing the relationship of teaching and, moreover, the relationship of knowledge. In this way, the author warns and observes ethics in research about possible interference by the subject (researcher) in relation to the object: *Ethics obliges us to take our distance and stay away from this "contact", because we would be "contaminated" by emotions and other feelings that would overshadow the objectivity of science.*

Self-analysis is treated as a "testimony" for "decontamination". Baitz ironically points out that nobody likes to be the object of analysis, not even science and scientists. The author cites the popular saying "in a blacksmith's house, the skewer is made of wood", comparing it to today's scientific world, noting that "not everyone is willing to bear the weight of their own truths". Or even so, *when the object is your own habitat, it is advisable that others also contribute in order to maintain scientific rigor, when it also implies letting yourself be analyzed, analyzing the most intimate motives of a research.* The relationships between the subject-object and a person's other institutional belongings, recalling the researcher's social mandate.

These are all presuppositions of the implication, they will give the "oneness" and set up a complex picture.

Baitz turns to Morin[4] :

> The key problem is not to reduce or separate, but to differentiate and bring together. The key problem is that of a thought that unites, which is why the word complexity (...) is so important, since complexus means "that which is woven together". Complex thinking is thinking that strives to unite, not in confusion, but by operating differentiations.

---

4 MORIN, Edgar. For a reform of thought in PENAVEGA, Alfredo and NASCIMENTO, Edgar Pinheiro do. Complex thinking: Edgar Morin and the crisis of modernity. Garamond, Rio de Janeiro, 1999, p. 33.

(MORIN, 1999 apud BAITZ, 2006 p. 32)

In the sub-theme "over-implication" or implication without limits, the author positions himself in favor of an evaluation that aims to achieve significant quality for the student. In the analysis of over-implication, the author states that *implication is not neutral, it is participatory, dialectical and refers to the researcher as an active subject in the field and is always presided over by more than one person, who is not always in academia, exploring the movement of participation and sharing of knowledge.*

The debates on the subject and object of this type of research in the field also include the reference to the social sciences in particular, as they have a very intimate relationship with the object. Added to this is the sociological perspective, which the author then summarizes as a "state of the art", in his words it seems to be something that encompasses all "atmospheres". And he organizes the object as the traverser of the subject. "By definition, the choice of profession implies full time".

In the pedagogical practices for an effort towards knowledge, when he quotes Paulo Freire, in which he places the relationship *educator - student - content,* Fieldwork is like a means for implication. Also part of the practice is the presentation of the analysis, as the author points out in Lefrebvre's work when he mentions what Baitz says is the most classic method: "After the analysis comes the exposition. If this is done successfully, the life of the object considered and the movement of the matter studied is reflected in the ideas presented. To such an extent that readers sometimes imagine that they are dealing with an a priori construction of the object."

The relationship between the researcher and the object, between the researcher and the institution, between the researcher and science (...). The rapprochement is made in order to combat the (non-existent) neutrality revealed as a way of achieving greater objectivity (BAITZ, 2006. p. 43).

The author invites us to reflect on hierarchy, dialectics as an indication of progress in the way of thinking and understanding the world does not mean that there is a hierarchy between thoughts. "History shows the opposite, because the movement of progress has also admitted retrogression, especially with regard to the adoption of dogmatic forms of understanding (...). In this way, the implication is projected against hierarchical forms, in the sense that the fragments show the whole, and what is banal and latent to the eye leads to complexity, slowly and towards the essential."

Finally, the author invites you to get involved, to reflect on your own implications and to practice them. It seems to invite a methodology of questioning and notes in fieldwork. Like a dialog between the subject and beyond the fields of vision that guide them. Although exhaustive, the concept of implication is interesting because it points to and highlights the need to search for other sciences. It also shows fieldwork as a participatory research methodology, suggesting an interesting relationship with the world, providing opportunities for searching and understanding through implication.

3 - Alentejano, Paulo R.R & Rocha-Leão, Otâvio M

Concepts:

Landscape evolution models; Field monitoring; scale, expedition/exposure

Topics for debate:

S Fieldwork and geographical theories; articulation between theory and practice

S The importance of fieldwork for geography research and teaching

S Fieldwork as a tool for geographers

S AGB as a disseminator of science based on fieldwork practices

S The dangers of trivializing fieldwork

S Theory and method

S Dichotomies in Geography

S Transformative action

Information Technology

- Alentejano, Paulo R. R. and Otavio Miguez da Rocha Leao (Rocha-Leâo, Otavio M.)

ALENTEJANO, Paulo Roberto Raposo; Otavio M. ROCHA-LEÂO. Fieldwork: an essential tool for geographers or a trivialized instrument? Boletim Paulista de Geografia, v. 84, p. 51-67, 2006.

In this article: "Fieldwork: an essential tool for geographers or a trivialized instrument?", geographer professors from the State University of Rio de Janeiro (UERJ) discuss the importance of fieldwork for geographers, as well as the problems surrounding this tool.

With its emblematic content on geography, the article is built on a "dialogue" between the professors. One a geomorphological geographer, the other a human geographer with a sociological bias, suggests subjects such as: fieldwork and theory; the importance of fieldwork for research and for teaching geography, and; the dangers of trivializing fieldwork.

They begin with a formulation of geography as a science and the role of fieldwork since its beginnings, and comment on the empiricist legacy of geography:

> (...) while this legacy was fundamental to the consolidation of Geography as a science, it also left a strong empiricist imprint. Thus, in the early days, fieldwork, which was a fundamental part of the method, gradually became the method itself, i.e., from being part of the method, it became the method, the result of the predominance of an empiricist conception that despised theory and attributed to the description of reality the condition of a criterion of truth. (ALENTEJANO & ROCHA-LEAO, 2006 p. 53)

Highlighting this dimension of empiricism present in Geography, (THOMAZ Jr.

2005 apud ALENTEJANO & ROCHA-LEâO, 2006 p. 53) argues that the main concepts formulated with regard to Fieldwork are: (1) "without field research no one has the right to speak" Mao Tse Tung; (2) "no more theory, the important thing is to do"; (3) the primacy of description over theoretical reflection; (4) a space for synthesis between - kept separate - physical and human elements.

Such conceptions reveal not only the dominant empiricism in Geography, but also the separation between a Geography of men and a Geography of nature, as if they were absolutely distinct realities. In this sense, it reveals the case of Brazilian geography and the permanence of geographic paradigms.

In this sense, Alentejano highlights "the contribution of the Association of Brazilian Geographers (AGB) to the dissemination of fieldwork as a fundamental tool for geographers, since its first congresses represented a privileged space for the production of Brazilian Geography, with emphasis on the excursions it promoted, the lectures it organized and the texts it published in the Boletim Geogràfico". (ALENTEJANO, 2002, p. 146 apud ALENTEJANO 2006, p. 54)

All this accumulation was thrown into disarray from the 1970s onwards, when, in the wake of the hegemony of Theoretical and Quantitative Geography, fieldwork was condemned and practically wiped off the map of geographers' practices, on the grounds that information technology and mathematical models would be more appropriate tools for investigating reality (ALENTEJANO & ROCHA-LEâO, 2006).

Also in the initial movement of critical geography, the radical criticism of the empiricism dominant in traditional geography led to a denial of the validity of fieldwork as an instrument for constructing geographical thought, due to the emphasis on theory. (ALENTEJANO & ROCHA-LEâO, 2006)

(...) It has become exaggerated to only value theoretical contributions with a Marxist foundation (...) and to neglect previous contributions and, above all, the empirical knowledge produced on the basis of fieldwork. Today, when the value of technique is once again emphasized, with the support of new tools (remote sensing, GIS), it must be stressed that these are only tools and not foundations. The difficult alliance between theory and instrumental practice must be pursued and always based on field research." (RUA, 1997, p. 45 apud ALENTEJANO & ROCHA- LEÂO, 2006. p. 55)

The difficulties of articulating theory and practice are intertwined with the problem of the dichotomy between Physical and Human Geography, because, despite the discourses that preach an *integrative vision* of Geography, in the sense of not isolating the physical and social variables responsible for the production of geographic space, the researcher's gaze, historically already impregnated with the physical-human dichotomy, ends up producing, more often than not, an eminently social or natural approach to the phenomena manifested on the earth's surface. (ALENTEJANO & ROCHA-LEÂO, 2006, p.55)

In this sense, with the exception of French Regional Geography at the beginning of the 20th century, field research in Geography has followed the path of specialization in phenomena of nature or society, and it is therefore necessary, albeit painful, to separate Physical and Human Geography, at least for an assessment of how field procedures have evolved so far, and how they have reflected in the production of fragmented knowledge of reality, through the strengthening of specialties. (ALENTEJANO & ROCHA-LEÂO, 2006. p.56)

As this was a dialogue between physical and human geographers, it was to be expected that these conclusions would lead in this direction. But it's important to note that, in the course of geographic thinking, the importance of the AGB in disseminating the results of the fieldwork carried out, despite having been interrupted, the current of geographic thinking based on the field tool is back on the pages of the AGB, albeit without the centrality it used to have. In addition to other tools that have come about as a result of technological advances that support geography.

In this sense, there is a bifurcation in the history of geographic thought, since both Human Geography and Physical Geography have followed the path of positivism and neopositivism, but the same has not been true of dialectical materialism and phenomenology, which have developed in Human Geography, but have had practically no penetration in Physical Geography, where, in turn, general systems theory has developed - which has had less expression in the Geosciences.

Another sub-theme that the author is concerned with answering is the relationship between theory and method. He points out that fieldwork is the moment when theory and method come together, and not just reduced to the empirical world. In this sense, the author cites practice-theory, in other words, fieldwork as a process of building knowledge, which, together with theory, will reveal the essence of geographical phenomena.

In this way, observing the landscape alone becomes unfeasible, as it cannot be reduced to this exercise alone. This is why geographic concepts are so important in mediating the dynamics that occur in space. This article also highlights the centrality of the concept of space, as well as the concepts that will mediate it, such as the articulation of scales.

Bernard Kaiser is cited for understanding fieldwork as a tool at the service of the geographer, as long as it is articulated with theory, capable of enabling a connection between empirics and theory, in other words, fieldwork is a means for the geographer and not an objective in itself.

We should also add to this article Fieldwork and its use for research purposes, highlighting method and theory, the use of technologies as tools for the method.

Furthermore, despite the obstacles and the paradigms of geography, the author supports the use of fieldwork as a research tool, as we have seen above, and for didactic purposes in geography teaching, highlighting in particular the problems that teachers face when opting for this tool. "At the various stops along the way, teachers make a speech in front of passive students", as Lacoste wisely put it (LACOSTE, 1985. p 13 apud ALENTEJANO & ROCHA-LEAO, 2006 p. 63).

It may be interesting to portray this moment, but other aspects should be considered when going out into the field with secondary and primary school students. These excursions, as they are called, if they are prepared in advance, can give students a problematizing look and prepare them for what they are going to observe. However, it is in the field that social and natural realities intersect. In this sense:

> (...) the development of field itineraries concerned with highlighting the natural and social phenomena (and especially the integration between them) that shape the earth's surface can become an important integrating tool in the training of new generations of geographers who are more attentive to the physical-human relationship, without necessarily neglecting the advance-verticalization of spatialities. (ALENTEJANO & ROCHA-LEâO, 2006 p. 64)

In other words, teaching also involves articulating scales of analysis, taking care not to fragment reality by sticking to the human and/or physical terrain, as well as implementing monitoring activities directly linked to physical geography.

In addition to these questions about research and teaching, fieldwork is once again gaining ground on the geography scene. In the course of the article, the choice of fieldwork with the aim of serving private accumulation has been warned against, but it has to be said that this is optional. Rigorous fieldwork, which escapes the crisis of being a trivialized instrument, needs to be committed to uncovering the mechanisms of construction of domination and exploitation and, above all, committed to transforming reality. At the same time:

> It is interesting for the results of field research to be passed on to the people, so fieldwork should be associated with two facets: the articulation between theory and practice; and a critical look at reality associated with transformative action. (ALENTEJANO & ROCHA- LEÂO, 2006 p. 64)

A few comments on this chapter show that by starting a dialogue between the authors, they are seeking to (re)bring the fields of geography closer together, provided, however, that they are willing to review theory and practice in order to build knowledge.

4 - Luis Antônio Bittar Venturi

Concepts:

"Thinking and doing", scientific work, technological instruments - "seduction of technology", laboratory, fieldwork, "office".

Topics for debate:

S Method and technique are complementary;

S Technical work vs. scientific work;

S Techniques evolve according to human needs;

S Mastery of technique = complexity of new technologies x option of using simpler instruments.

S Laboratory techniques

S Fieldwork techniques = flexible planning.

S Specific objects of study for specific techniques;

- Luis Antonio Bittar Venturi

VENTURI, Luis Antônio. The role of technique in the process of scientific production. In: Boletim Paulista de Geografia, v. 84, p. 69-76, 2006.

We will begin our analysis of Venturi's text by observing *the role of technique in the process of building knowledge*, since "the sciences that work on the most different aspects of the real world, (...) techniques play an important role in the process of scientific production", noting that method and technique and instruments are complementary, or even that the use of technique makes it possible to obtain data on reality that will support the paths taken by the method.

In addition to these statements, the author comments on the technique used in scientific work (differentiating technical work from scientific work), guaranteeing method a privileged place in the problematization of research. For the researcher, Venturini warns of the need to be aware of the significance of technique and its role in the process of scientific production, for which a real mastery of technique must be required.

Driven by the market, technological advances enable and create possibilities for techniques to evolve. These are driven by the needs of their time and gain complexity as they unveil others. It is worth admitting that man creates his techniques for his own needs and puts himself in front of a game of *technological seduction* (albeit unconsciously).

> In the logic of the market, many instruments (such as various software and satellites) and the techniques associated with them are developing with astonishing speed. Technology developed to meet the needs of productivity and profit undeniably brings many benefits to humanity in various areas (...). New technological discoveries can meet both real needs and those created artificially to maintain market dynamics (cell phones with games, increasingly powerful digital cameras, etc.). However, if access is restricted by inequitable purchasing power. Although there is talk of the massification of technology, paradoxically, access to it is not universal. (VENTURI, 2006, p. 71)

Furthermore, the process of scientific production consists of three stages: 1.

Choice of technique; 2. Cost-benefit; 3. Feasibility. This means that using a current technique requires complex handling, which, if you choose a simpler technique, ends up giving the same result with less complexity. Solution techniques are the choice of simple techniques for a process that meets items 2 and 3.

At the end of the day, the author differentiates the laboratory from the field, solving the equation of producing science in the following way: In fieldwork, planning must be flexible, while in the laboratory, forecasting is more static. In any case, the use of the specific technique for the specific object is as important in field work as it is in laboratories.

5 - Yves Lacoste

Concepts:

Responsibility of the researcher; research results; expedition/exhibition; subject-object of research; landscape; evocation of landscape; man-nature relationship;

Topics for debate:

S Research and Fieldwork: political problems;

S The role of the university; - collective research vs. individual research;

S The social role of the researcher; - proposing solutions/communicating the results, what will it provoke in the group studied?

S Commitment to the organization;

S Theoretical training + practice + articulation of scale;

S The "power" of research;

S The role of the state and big business;

S The technology monopoly;

S Budgetary difficulties;

S Shipping/Exhibition

- Yves Lacoste:

LACOSTE, Yves. Research and fieldwork: a political problem for researchers, students and citizens. Herodote Magazine n° 8, Oct./Dec. pages 3 to 20, 1977. In: Boletim Paulista de Geografia, v. 84, p. 77-92, 2006.

"Research and fieldwork: a political problem for researchers, students and citizens", is a text originally published in French in 1977 by the magazine Herodote (n° 8, Oct./Dec. pages 3 to 20) in which Yves Lacoste argues about problems of field research that are rarely evoked, perhaps because they are fundamental.

A structure is needed for the construction of knowledge, where the elements are recognized and named according to a problem and inherent assumptions. "Above all, we need to **pose political problems**, and first and foremost the responsibility of the researcher towards the men and women he studies and whose territory he analyzes." What is evident, however, is the multiplication of research of all kinds, a factor that implies countless scientific publications, which favor the state and large private groups to elaborate and execute projects that, for reasons of the impossibility of reviewing all the material, are rendered vain.

By illustrating this situation, Lacoste leads us to reflect on the paradigms of geography, "which, even when research focuses on the physical characteristics of a territory, do not fail to say less about the people who live there, to the extent that their results make intervention there more effective". (LACOSTE, 2006, p.77)

On the subject of Pierre Gorou and the beauties of the Delta, Lacoste seeks to elucidate issues that have been criticized in relation to the "evocation of the landscape" of the Delta, which dealt with lengthy descriptions of the landscape. The answers to the criticisms came from the organization of the facts and the factor of hiding the misery and later the wars that took place in this locality.

Regarding the concept of landscape, Lacoste put the beauties of the Delta (text) as an unconscious way for Pierre Gorou to distance himself from the problem of misery and hide it, as he didn't comment in the text on the repression and terrible famine that the population had gone through. However, he had written that the Delta was an example of the *perfect harmony between man and nature,* also known as the "harmony of landscapes".

Moreover, with greater reason, the population will ignore all the results that have been submitted to them. For this reason, Lacoste insists that it is necessary to take responsibility towards the population that has been studied "in the effort to communicate the results to the men and women who have been the objects of the research, because these results confer power on those who hold them."

As for **the problem of the researcher's responsibility** and the relationship that should be established between the intellectual and the population that is his scientific center. As Lacoste says: it's men and women we take sides with, translated as the object of research. That's why it's necessary to establish a researcher-researched relationship, from which political problems are imported, so that experiences can be confronted.

In addition to the logic of power, there is also a monopoly on new information technologies and the archiving of data using computers, with the state funding investment in this research. With regard to universities, there is criticism of the budgetary difficulties in promoting the production of technologies and organizing field research.

Another problem at university is the student experience in research, which is limited to being an object of reproduction. What is important, according to Lacoste, is the organization of the research and the commitment to responsibility in the subject-object relationship. Both in collective research (usually associated with undergraduate research) and in individual research (master's degree).

To this end, Lacoste proposes a renewal of thinking for a democratic transformation of the field research relationship, organizing the "expedition/exhibition" relationship. It is a question of sharing theoretical training, practices, which are insufficient if not worked on in harmony, and articulating the scale of the question and what this study will provoke in the group studied. In this way, problems are posed that affect them directly and, for Lacoste, the results obtained have power in the hands of those who hold them, which is why it is important to present the results.

6- Bernard Kaiser

Concepts:

Sub-system; situational analysis, local analysis, historical analysis; local/global

Topics for debate:

S The importance of field research;

SPesearcher = citizen;

S Typology of fieldwork for the social survey;

Specific geographical field research?

s Methodology proposal: bibliographical references of the site + theoretical background + repercussion of the hypothesis;

Social proposal: conflicts and problems; class struggles;

s Everyday life = the indecipherable;

s Justification of the research and the researcher's responsibility;

s Valuing the use of local people's knowledge;

- Bernard Kaiser:

KAISER, Bernard. The implication: a new sediment to explore in Geography? In: Boletim Paulista de Geografia, v. 84, p. 25-50, 2006.

In *The geographer and field research*, Bernard Kaiser proposes a re-reading of the article published by the author in the magazine HERODOTE n° IX, under the title: Sans Enquête, pas de iroit à la parole. Translated from the original French by Antonia D. Erdens and published in Seleçao de Textos n° 11.

The article looks at the foundations, methods and practices for social survey research. Firstly, the question of fieldwork as an obligatory tool for the citizen, guided by the idea that the *researcher does not stand apart from the man, from the citizen, he is often a teacher at the same time, which multiplies his responsibilities.*

Is there really such a thing as specific geographic field research? In response, the text states: "in fact, the objectives are much broader, (...) because it is a question of discovering, within its complexity and globality, the reality of a localized social subsystem". "If the survey wants to get to the heart of reality in order to collect the elements necessary for analysis and explanation, it must penetrate the forces and relations of production, explore the ideological and cultural levels of social dynamics."

This is not a methodological recipe. Situation analysis is a project, and is the fundamental method for getting to know a situation. From a social point of view, the situation is complex, conditioned by interpenetrating processes and mechanisms whose in-depth study requires the exploration of all economic, cultural, political and ideological aspects, both in the present and in the past.

How to do this analysis. According to Kaiser, the frame of reference for local analysis. It is research that is indispensable for the analysis of the social situation, and he sees field research at the service of the people as a justification. The social situation is first and foremost a product of history, of class struggles and the way in which they are translated on the ground. The most diverse actors are involved in these processes: social groups, state apparatuses, institutions, the media and ideologies.

Finally, for Kaiser, the vision is social and not spatial. Space, according to him, cannot be studied by geographers as an independent category, but rather as a strategy, adding to the local/global articulation.

7- Valeria de Marcos

Concepts:

Participatory research; community production; teaching-learning process, educational action, totality.

Topics for debate:

S Discusses the importance of fieldwork in geography;

S Participatory research as a methodology and experience;

S The way participant research is carried out;

S The purpose of the research and the results it could leave the communities studied;

S The attitude of the researcher when approaching the community - "The arrival and stay of the stranger" - ;

S The difference in temporal understanding for the researcher and the community being studied; local vision;

S The importance of the gaze in conducting research;

- Valéria de Marcos:

MARCOS, V. Fieldwork in Geography: reflections on an experience of participant research.

Boletim Paulista de Geografia, v. 1 n 84, p. 105-136, 2006.

This article discusses the importance of fieldwork in geography, based on an experience in the Japanese community (located in Sao Paulo). It is a discussion of participatory fieldwork in geography, especially the ways of carrying out participatory research, involving the researcher's commitment to the community and the researcher's attitude during the approach and stay in the community. "Thinking about fieldwork as a technique and bringing up questions that should guide our practice and attitude as researchers."

Arriving in the study area is a delicate moment in the article, in which the author tries to elucidate her experience in relation to the presentation and her (com)posture so that there is a link between the researcher and the community. The moment of acceptance by the community becomes fundamental at the beginning of the research, because it is in this way that details that seem functional in the final analysis are achieved. From the outset, the researcher must maintain a foreign gaze, accepting the dynamics and, in a way, allowing them to participate in order to contribute to the study.

The importance of fieldwork in Valéria de Marcos' view has already been communicated as a didactic and research tool of fundamental importance for teaching and research in geography.

> As a didactic resource: it's the moment when the student visualizes everything they've seen in class, when theory becomes reality, when it "materializes". I would also say that planning is a moment of caring about the production of knowledge, in other words, the teaching-learning relationship. As a research tool: it is one of the most important for the construction of geographical knowledge, the moment when the theme becomes a reality, exercising the look with depth and observation beyond what had been considered. (MARCOS, 2006, p. 106)

The discussion is also based on an analysis of the time dimension of the researcher in relation to the community they are studying, their approach to carrying out the research, the choice of the purpose of the research and the expected results. These factors are the attributes that characterize participant research. In addition, the role of the researcher is one of the reflections that are appropriate when preparing to go into the field.

Going into the field means participating in the research (getting to know the people being researched better), so despite the planning, the approach must be flexible, and there may be the addition or extraction of content.

However, the question of ethics in research is raised: What should be done with the data collected? As a way of answering the question, we can use the following concept: educational action, which Valéria de Marcos assimilates with the knowledge produced in the research, must be given back to the community, so as to allow them, at the same time, to increase their awareness of themselves and their capacity for transformative initiative.

This article also raises questions about theory, practice and method. It draws attention to the problems posed by the experience in a community of peasants, extracting the possibility of visualizing another geographical concept, the relationship between countryside and city, showing itself as a "passport" for cultural research.

2.2. Analysis of references

- Table of references:

References cited in articles	Occurrences
YVES LACOSTE (1,2,3,4,5,6)	6
MAO TSÉ TUNG (3,6,7)	3
BERNARD KAISER (3,4,6)	3
MILTON SANTOS (1;4)	2
HENRI LEFBVRE (1,2)	2
PAULO FREIRE (2)	1
ANTÔNIO CHRISTOFOLETTI	1
PAUL CLAVAL (1)	1

Note: The references are organized from the most referenced and in the sequence offered by the BPG.

In view of this table, we should consider the quotes from Yves Lacoste, which focus on a debate about the *political responsibility of the researcher:* "Above all, it is necessary to pose political problems, and first of all, the responsibility of the researcher towards the men and women he studies and whose territory he analyzes" (LACOSTE, 2006, p. 77); *the power of research:*

"Is it normal for a researcher to be disinterested in the fate of the population he has studied? Is it normal for them to remain ignorant of the research of which they have been the object? This problem, of the researcher's responsibility towards the men and women he studies and whose territory he analyzes, was raised by Hérodote in its first issue, and at least as far as I know, we are still almost the only ones to say that it is necessary not to stop the research, a negative and perfectly unrealistic attitude, but to make an effort to communicate the results to the men and women who were the objects of them, because these results give power to those who hold them." (LACOSTE, 2006, p. 78)

He also questions *the role of the university,* contemplating the discussion of *collective research vs. individual research; the social role of the researcher; proposing solutions/communicating the results,* under the question*: "what will it provoke in the group studied?";* Field work must be *committed to the organization;* in addition to *theoretical training + practice + articulation* of *scale;* Lacoste also comments on the *role of the state and large companies and the monopoly of* *technologies.*

This has led to a debate on fieldwork based on Yves Lacoste's original publication and several authors have chosen to use it as a reference for research on the subject of fieldwork in geography.

Milton Santos, the quotes were about the relationship of space in the format of Fieldwork. This is one of the "facets" in which the geographer is constantly referenced, among others, in terms of understanding the relationship and interaction of the "elements" of space, the notion of "cause-effect", "functional spatialization" and "true totality", the latter contained in Henri Lefbvre, also cited as a reference for teaching geography.

Bernard Kaiser is also a reference, when the authors rely on the action of the research subject and the valorization of local knowledge in research. Mao Zedong made his presence felt with the phrase: "Without fieldwork, no one has the right to speak." Other discussants will be called upon throughout the text, such as Antônio Christofoletti (see book: Perspectivas da Geografia, DIFEL, 1985 - Chapter 2. Paul Vidal de La Blache The characteristics of Geography; Chapter 7. Yi-Fu Tuan, Humanistic Geography and Chapter 8. Apprehending the Dynamism of the Lived World - Anne Buttimer), Paul Claval (Cultural Geography) and Paulo Freire (Fieldwork and Pedagogy - methods of synthesis in the subject-object relationship), as they are complementary readings.

2.3. Chapter considerations:

The content of the Bulletin reveals the theoretical-methodological reflection, introduced by the authors with different nomenclatures in the analysis, but which intend the same reading such as: theory and methodology, theory and method, Fieldwork and geographic theories (referring to the history of geographic thought), these terms behave as the "translation" of Fieldwork:

> Gradually, this practice was assimilated, especially because - and this is all the more important - it was discovered that it did not break with science, but brought to light a new objectivity; going into the field and "feeling it" made research more complex and complete, as well as raising questions that had been obscured until then. It was, in short, a magnificent advance, an updating of science (BAITZ, 2006, p. 29).

Among the articles, the findings regarding fieldwork and its importance for geography are both relevant to the construction of geographic knowledge and the teaching of geography, but also in terms of the researcher's responsibility and the revelation of the object through different methods, experiences, methodologies and worldviews.

We can then select six subsets of analysis that will mediate and construct the importance of Fieldwork in Geography (including itself as a subset), based on a review of the concepts and themes present in the articles in the Boletim Paulista de Geografia (BPG, NÙMERO 84. SP, JUL.2006 - O Trabalho de Campo) and based on an analysis of the summary table (Appendix 1 - Summary Table

- Complete):

- The importance of fieldwork for geography;
- Fieldwork and the construction of geographical knowledge;
- The subject-object relationship;
- The "place" of fieldwork and the "scale";
- Inserting technology into field research.
- Theory and Practice

Ângelo Serpa, the first article in the BPG, discusses Fieldwork as a specific discipline of Geography. The author points out that "a significant spatial cut-out conceptualizes the geographer in his objective for spatial analysis". (...) It is through Fieldwork that the geographer organizes "the various possibilities of cutting out, analyzing and conceptualizing space, according to the questions, goals and objectives defined by the research subject" and that Bernard Kaiser contributes with differentiations:

> Field research is a means and not a goal in itself. It is research that is indispensable for analyzing the social situation. It is, we repeat, a social situation and not a spatial situation. Space cannot be studied by geographers as an independent category since it is nothing more than one of the elements of the social system. (KAISER, 2006, p. 58)

The importance of fieldwork, in Valéria de Marcos' view, has already been communicated as a didactic and research tool of fundamental importance for teaching and research in geography. In the articles in Boletim Paulista, the authors converge on this thought. Fieldwork is an instrument that makes Geography "efficient" and present in all the objects it sets out to analyze:

> Although contested, the practice of subject-object separation has unfortunately taken deep root. Paulo Freire, to cite only the field of education, pleaded for the extinction of *banking education*; in other words, The pointing out of progress in the way of thinking and understanding the world does not mean that there is a hierarchy between thoughts. (...) He intensely declared that the teacher and/or knowledge are not subjects vis-à-vis the students (who in turn are not objects), but that both are in relationship, going beyond the analytical method in education. Despite this, the teacher who declares himself to be a learner in the classroom today continues to be viewed with suspicion of incompetence and incapacity by students, parents and professional colleagues alike... "Teacher and student don't mix", is what they say, hierarchizing relationships and knowledge itself, backing themselves up and putting Piaget's theory to a use that would sadden them (BAITZ, 2006, p. 27).

When it comes to the researcher, there is a clear debate about the problematization of subject and object, which evolves into the "non-neutrality" of science:

> Another issue that had also become clear was the discussion about the neutrality of science. I finished the course convinced that there is no such thing as neutral science, because there is no such thing as science without commitment. In fact, the choice of research topic and who will help us along this road reveals, at the same time, the commitment and

the science we intend to do. (MARCOS, 2006 p. 107)

For example, "the arrival and stay in the study area", in Valéria de Marcos' article, is the way in which the author sought to elucidate her experience in relation to the presentation and her (com)posture showing the subject-object relationship. She also says that she has to look for the "link" between the researcher and the community. The moment of acceptance by the community is fundamental at the beginning of the research, as this is how details that seem functional in the final analysis are achieved. From the outset, the researcher must maintain a "foreign gaze", accepting the dynamics and, in a way, allowing them to participate in order to contribute to the study.

Before going into the field, it's important to emphasize how important the method is for analyzing the importance of fieldwork and suspending the synthesis between subject and object that follows in Serpa's conclusions:

As methods (dialectics and phenomenology) they can work as complementary strategies, always seeking to build the subject-object synthesis that is inherent to the act of knowing, sometimes using history as a category of analysis, sometimes intentionally seeking to abstract the historicity of phenomena, aiming to make their "essence" explicit. (SERPA, 2006, p. 20)

A concept that is pertinent to the discussion of geographical knowledge is the diversified subject-object relationships, noting the use of specific techniques for specific objects (VENTURI, 2006, p.76). This topic opens up a debate on the dichotomies between Physical Geography and Human Geography and the methods that stem from them. Even though there are different objectives, it is worth reflecting on the unification or better use of general fieldwork situations in the context of research for, and only for, geographical knowledge.

Fieldwork should be based on the totality of space, without forgetting the specific arrangements that make each place, city, neighborhood or region a particular articulation of physical and human factors in a fragmented but (increasingly) articulated world. Fieldwork in Geography must therefore pursue the idea of particularity in totality, emphatically abandoning the idea of the singularity of places, cities, neighborhoods or regions. (SERPA, 2006, p.10)

Valéria de Marcos mentions some initial concerns about carrying out fieldwork. She draws attention to the relationship with "place":

How do we "get there"? How do we approach people? How do we win their trust? How do we guarantee the scientificity required by academia? These and so many other doubts that guide us during research often mean that the topic itself takes a back seat at this early stage. The choice of how to carry out the research will depend on the subject and the objectives of the research that we have to carry out. (MARCOS, 2006, p. 106)

We must understand from this knowledge that *scale* is a relevant concept in the analysis because we will be dealing with a set of "sub-systems" (KAISER, 2006) or also by Ângelo Serpa, who proposes scale as a *mediator in* the analysis of space, and who approaches geographic concepts up to the extension to places.

Pesquisaclor - Scale → Observation unit and attributes

According to this concept of "sub-system", we can get closer to the concept of "sub-culture" emerging in discussions of Cultural Geography and especially in the contemporary social scene. In Bernard Kaiser's analysis, this is revealed through the concept of situational analysis:

> The analysis of the situation must take everything into account: in essence, this is what is known today as a system analysis. The local situation is, in reality, a sub-system of a meta-system representing the social formation. It is therefore necessary to understand it in systemic terms, rejecting the inventory of determinants - the usual geographical work - and the Cartesian study of structures in order to go straight to the workings, to the processes. (KAISER, 2006)

Lacoste, for his part, defines fieldwork as large-scale research that must necessarily be linked to other scales: "If fieldwork is not to be merely empiricism, it must be linked to theoretical training, which is also indispensable. Knowing how to think about space is not just about placing problems in a local context; it is also about effectively linking them to phenomena that develop over much wider scales." (apud ALENTEJANO, 2006, p. 58)

For this author, therefore, the field is only valid if it is articulated with global systems for interpreting reality. It is through the articulation of scales that we can effectively construct a geographical interpretation of reality, going from the particular to the general and back again, as well as from practice to theory and vice versa.

Another point that underpins the discussions is the introduction of new technologies into the format of fieldwork, which other authors have described as the "seduction of technology" (VENTURI, 2006) and the dubious universality of technologies applied to geography:

> With the exception of a few works, field research is a presupposition in geography. Transcription (description), representation (cartography) and modification (planning) are all activities carried out by geographers that involve working in the field at some point. That's why it's possible to say that we developed an almost tactile intimacy with the field during our training, and that it's a great pleasure for us to go out into the field. This intimacy, however, doesn't make us totally free in our research activities, because we don't just go out into the field. The "going" is preceded by many techniques and technologies. We carry certain tools, such as a research diary, and always knowledge of the techniques and procedures of analysis. (BAITZ, 2006, p. 26)

> In the logic of the market, many instruments (such as various software and satellites) and the techniques associated with them are developing with astonishing speed. Technology developed to meet the needs of productivity and profit undeniably brings many benefits to humanity in various areas, including medicine, communications, etc. New technological discoveries can meet both real needs and those created artificially to maintain market dynamics (cell phones with games, increasingly powerful digital cameras, etc.). However, their access is restricted by unequal purchasing power. Although there is talk of the massification of technology, paradoxically, access to it is not universal. (VENTURI, 2006, p 71)

However, it is necessary to understand technological progress as a trend and a possibility. The development of new information technologies will not make it unnecessary to carry out fieldwork, especially in order to obtain data that dialogues with reality and according to the geographer's needs:

> New technologies make the field easier, but without method there is no production of knowledge, after all, instruments are means of working. According to this author, research is seeking answers to thought-provoking questions, in a process in which subject and object interact, the subject constructing the object and the object reconstructing the subject. (SUERTEGARAY 2002, apud ALENTEJANO, 2006, P. 59)

CHAPTER 3

Fieldwork at University and School

We should first point out the different objectives of the academy and the school: the former aims to provide higher-level professional training, while the latter aims to provide citizen training based on specific knowledge of the subject. The subject of school geography is not, however, the result of a simple *didactic transposition, a simplification of academic knowledge into school knowledge* (PONTUSCKA 2009), but is based on the science of reference, so that it itself becomes a means of constructing knowledge.

3.1. Fieldwork at the FEDERAL UNIVERSITY OF ALFENAS

In order to understand these two fields of teaching, the Federal University of Alfenas proposes competencies and skills for the training of teachers at UNIFAL, including "Planning and carrying out field activities related to geographic and environmental research"[5] ; for us future teachers, this is an essential training activity.

Fieldwork at UNIFAL is organized on an individual basis, such as Geomorphology Fieldwork - to Nazareno - SP and also to Campos do Jordao - SP (Professor responsible for the subject: Marta Felicia Marujo Ferreira) and also with the same professor to Ubatuba - SP with the subject of Coastal Geomorphology. To the Minas triangle, with the subject of Regional Geography (Teacher responsible for the subject: Ana Rute do Vale) and Sao Joao Del Rey and Tiradentes with the subject of Formation of the Brazilian Territory (Teacher responsible for the subject: Ana Rute do Vale) and to Sao Paulo with the subject of Urban Planning and Urban Geography (Teacher responsible for the subject: Evânio dos Santos Branquinho). As well as other subjects that apply Fieldwork in the Alfenas - MG region itself, in order to explore the region and the geographical phenomena that characterize it.

But integrated fieldwork has also taken place more frequently in recent years. The latter took place with the subjects of Urban Geography (Teacher responsible for the subject: Evânio dos Santos Branquinho) and Economic Geography (Teacher responsible for the subject: Flamarion Dutra Alves). And also the Field Work to Prado - BA, in which a debate was organized around Regional Geography (Professor responsible for the subject: Ana Rute do Vale) and Natural Resources of Brazil (Professor responsible for the subject: Clibson Alves dos Santos) (this practice is enriching, adds Ana Rute), in addition to other practices such as Field Work, which thus demonstrate the

5 Available at PROJETO POLÌTICO PEDAGÒGICO DO CURSO DE GEOGRAFIA - licenciatura 2011. <http://www.unifal-mg.edu.br/graduacao/?q=geolic> and on the UNIFAL-MG Geography website that follows:< http://www.unifal-mg.edu.br/geografia/?q=node/3>

presence of UNIFAL geographers in various places in Brazil.

In this way, the aim is to verify, through the questionnaires applied, the relationship between theory and the academic practices carried out by the Professors of the Institute of Natural Sciences - ICN - Geography of the Federal University of Alfenas. They were asked about the places they research and for which subject? Have you ever worked with other subjects in your fieldwork? (described above) In your fieldwork, is there an emphasis on exploring the socio-cultural reality of the place? What are the successes and limitations of your fieldwork(s)?

Among the issues, we can group them into the following topics:

- **Academic practice and theoretical study:**

In this first topic, we can verify the relevance of Fieldwork as a (disciplinary) objective of Geography at UNIFAL. There is a lot of discussion about these practices (integrative or not), and that in Geography, there are still conflicting debates in the constitution of the science itself, but it is undeniable that this is a necessity for the discipline since they are part of the visualization of academic practice:

> Firstly, we defend the idea that there is a disciplinary specificity to the discussion proposed here, specific to Geography and the production of geographical knowledge. A reflection on the importance of fieldwork in this discipline requires an understanding of its specificity in relation to other disciplines, of its strengths and handicaps in relation to other natural and social sciences. After all, what is epistemologically different about the production of geographical knowledge? (SERPA, 2006. p. 9)

Despite this assumption, in UNIFAL Geography each research group formalizes itself by expressing results that are less and less articulated with the reality of the Bachelor and Graduate Geographer, which is a point worth mentioning. Observing the words of Ana Rute do Vale, in which she mentions that "during their undergraduate studies, many undergraduates don't take part in practical work", implying a mismatch between theory and practice in the training of these professionals.

Fieldwork in Geography, according to Marta Ferreira: "should be the time to study, within a sequence that the teacher has proposed for that subject so that the student can reach the totality, the whole".

Professor Marta comments that the fieldwork in her subject (geomorphology) aims to build a notion of a "trail" from Alfenas to the chosen point, thus seeking an understanding of the dynamics along the route.

Professor Evânio Branquinho understands that in order to be in places, you need to know them beforehand, which is why his practice involves the presence of guests and/or the locals themselves, or the students themselves who have experienced the place visited, in order to provide "more open" notions for Fieldwork.

- Successes and limits:

1. Theory and Practice:

The teachers questioned praised the possibility of visualizing theory in practice. This is because when organizing fieldwork for their subject, or when they choose to integrate it, the teacher is already facing one of the faces of this methodology, theory in the classroom. Practice, being compulsory in the subjects, "carries" a sense of observing phenomena and it is in the realization of the field reports that the joining of theory and practice takes place in a clearer way.

In many situations, "the uncritical use of the label impedes" the theoretical deepening of this pedagogical practice which, if reduced to a visit, a walk or a field lesson, loses, from the perspective we are defending here, a large part of its formative and educational value (KAISER, 2006).

2. Budgetary and bureaucratic difficulties;

The professors at UNIFAL-MG also unanimously pointed to one of the major "obstacles" to carrying out fieldwork: budgetary difficulties. Similar to what Lacoste points out: "But even these excursions are becoming increasingly rare in most universities, certainly because of the serious budgetary difficulties (...)" (LACOSTE, 2006) according to the professors, this issue becomes even greater when the university (rectory) neglects the field experience as the main source of research for Geography:

> For the majority of students, the experience of research is limited to this; to this supplementary exercise of reproducing knowledge that has already been produced, without them being aware of the possibilities they have of producing elements of new knowledge on their own. However, learning how to do fieldwork, getting started with research, is beginning to become one of the main demands of Geography students. (LACOSTE, 2006, p.)

The latent concern over the financial issue is orchestrated by the length of time spent in the place, which means that the UNIFAL-MG Geography fieldwork doesn't last more than four days, except for the fieldwork to Prado in Bahia and to the Pantanal with the Biogeography subject. This evidence contributes to the demand for costs and the feasibility of cost-effective techniques. Even though the university provides the transportation service, there is still a need for demands, exclusive to geography, which will enable a significant improvement in the condition of the fieldwork.

- Socio-cultural reality in the background?

It is understood that the socio-cultural reality of the place visited is pre-eminent in Fieldwork, in other words, it is "there" even if it is not mentioned. For example, Professor Marta Felicio comments that "the objectives of fieldwork are geared to the discipline, but you can't turn a blind eye to this reality". Ana Rute replies that "even in the 'background', socio-cultural reality emerges". For Evânio: In Fieldwork, the student has the option of having several different visions, to build a vision, even if it's quick (because Fieldwork is quick) and generic (...), a vision of that place, of the problems of that place, because otherwise it would be a work with gaps, just the visualization of the material and the landscape without going into the social relations.

3.2. Fieldwork and geography teaching

The principle that "school is life": The problem of the relationship between SCHOOL and LIFE has been raised by educators in both Europe and America. More than once in this very plenary session, the principle that SCHOOL IS LIFE and cannot be closed off from what constitutes, at any given moment, the very context in which man makes sense, has been affirmed as a peaceful truth; it can no longer be a closed compartment, pseudo-preparing for life, outside of life. The Study of the Environment would be a "resource par excellence" for "[...] removing the boundaries between school and life" (MAGALDI, 1965, p. 69 and 70 apud PONTUSCHKA, Nidia N. 2009. p 5).

One issue that proved to be inspiring in the BPG is the use of fieldwork in geography teaching. Given its relevance and significance to the training of teachers at the time, the articles pointed to: 1. the construction of geographical knowledge through fieldwork; 2. overcoming the teaching-learning relationship; 3. the choice of a theme that is significant to the community, as ways of making better use of this practice.

As a didactic resource, fieldwork is the moment when we can visualize everything that has been discussed in the classroom, when theory becomes reality, when it "materializes" before the astonished eyes of the students, hence the importance of planning it as much as possible, so that it doesn't become a "recreational excursion" over the territory, but can be an extra moment in the teaching/learning/production of knowledge process. MARCOS

To a certain extent, the studies by Claudivan S. LOPES and Nidia N PONTUSCHKA: *Estudo do meio: teoria e pràtica (Study of the environment: theory and practice)* served almost entirely to reflect on and understand the historical moments of the Brazilian School and thus to analyze the theme of Fieldwork in the Teaching (and its curricular interpretation in the teaching) of Geography. According to the authors, the "study of the environment" is not a new pedagogical practice in the Brazilian educational universe:

It is actually part of a "school tradition" and is inspired by educators such as Francisco Ferrer y Guardia (1859-1909) and Célestin Freinet (1896-1966). Its aim is to provide students with learning that is "closer to life", in other words, more direct contact with the reality being studied, be it natural or social. (LOPES, PONTUSCHKA, 2009)

One of the fundamental stages of Environmental Studies is Fieldwork. The itinerary and timetable for the activities to be carried out during the field research should contain supporting texts and maps, a research itinerary, space for notes, texts and sketches, without going so far as to make it a recipe:

Field research reveals life, in other words, it aims to gain a more systematic understanding of how men and women in a given space and time organize their existence, to understand their needs, their desires, their struggles with victories and failures. Thus, during fieldwork, educators and students must immerse themselves in the daily life of the space to be researched, seeking to establish a rich dialogue with the space and, as researchers, with themselves. This is the moment to discover that the environment or space, in the interrelationship of natural and social processes, is a living Geography (PONTUSCHKA, 2006 apud 2009. P 14).

According to the authors, we still need to go into the field:

[...] without pre-judgments or prejudices: freeing up the way we look, smell, hear, feel, taste. In short, freeing up the feeling mechanized by life in society, for the affective reading that takes place in two opposing movements - denying alienation, the schema, the routine, the system, prejudice - and affirming the affection of community and personality (PONTUSCHKA, 2006, p. 12). By breaking the boundaries of the institutionalized territories of learning - the classroom and the school - field research allows this territory to be expanded, taking "the classroom and the school" out into the world - a more specific or particular place or situation in this world to be researched and studied - and the world - more real or concrete - into the classroom and the school. It is therefore an opportunity, as Thompson (1998) states, speaking more specifically of fieldwork in oral history, to generate learning opportunities beyond their traditional institutional shelters. (PONTUSCHKA, 2009. p. 15).

In this sense, we can make a connection with what is produced in Participatory Fieldwork or participatory research, presented by Valéria de Marcos, in which the importance of Fieldwork is already communicated as a didactic and research instrument of fundamental importance for teaching because, as a didactic resource, it is the moment when the student visualizes everything that was seen in the classroom, when theory becomes reality, when it "materializes". The planning of fieldwork also "supports" the production of knowledge, in other words, the teaching-learning relationship becomes more tangible and flexible.

Although there is a methodological procedure, fieldwork is a *flexible practice* (MARCOS, 2006), and success depends on flexible planning (but certainly rigorous, at least in the stages or moments). The authors point out that the following roadmap should not be seen as a denial of other organizational possibilities, but rather as a certain "strategic scheme", based on the authors' accumulated experience on the subject.

All the stages and respective actions that structure it are carried out in the search for possible pedagogical agreements and contracts which, without denying the conflicts that are consubstantial to any social relationship, have, as their starting and ending point, the reality experienced by the people involved in building an educational project in a given school unit. (LOPES, PONTUSCHKA, 2009. P. 7)

Beginning a process of modifications to this practice at specific moments in Brazil's political and educational history, the Study of the Environment (taking fieldwork into account as a fundamental part of this methodology) spread and became popular in the 1960s within the Escola Nova movement and among educators concerned with creating attractive teaching and meaningful learning.

Even so, there are records showing that it was carried out in schools founded by groups of European immigrants (early 20th century). Their arrival demonstrates the growing industrialization that was opening up jobs at the time, and that, inspired by Ferrer's pedagogical ideas, the schools created by the militants of the anarchist movement had the principle of offering rational teaching, based on field observations, discussions and the formation of a critical spirit about the surrounding environment. (XX)

Advances in understanding, the intensification of censorship and political repression promoted by the military government through Institutional Act No. 5 (AI-5), meant that Middle Studies were banned and with the crisis of the military government, from 1978-1979, and the consequent process of re-democratization of the country, "Middle Studies" returned to the agenda of educators. (LOPES , PONTUSCHKA, 2009)

According to the author, the return to the agendas was inspired by the pedagogy of Paulo Freire and his experience as Secretary of Education (1989-1990) in which the Study of the Environment played the role of an integrating link in interdisciplinary practices within primary school with very positive results, (PONTUSCHKA 2004, p. 260-268 apud LOPES, PONTUSCKA, 2009).

Recent research into *environmental studies* has shown that they are significant and up-to-date. However, it must be added that the indiscriminate use of experiences outside the school environment, even if they have pedagogical and recreational value, can be an obstacle to deepening the subject and expanding its meaning and potential in the development of the school curriculum. Let's take a look at the graph:

GRAPH 2

Nowadays, according to the research highlighted by the author, the "Study of the Environment" has contributed to strengthening the autonomy of the school institution and teachers in general, as Pontuschka (2009) points out:

From the school institution because it is an alternative to the policies and proposals coming from the education departments and from the teachers because they can develop their professional knowledge without being guided by official teaching materials. They can undoubtedly support the process of developing teaching professionalism. The **public dimension** of education: This involves organizing forums to discuss the problems experienced collectively by the community in which the school is located. By uncovering social injustices and other issues that affect the well-being of a given school community, Environmental Studies can be valuable tools for understanding and overcoming those injustices which, according to Santos (1993 apud PONTUSCHKA, 2009. P.6), "[...] must be corrected in the name of citizenship" (p. 112). We can thus speak of a school that is "rooted in place" and feeds off its potential and strength.

Among the proposals for this type of activity, Fieldwork can be understood in teacher training through the skills offered at UNIFAL-MG, through which these trained professionals will be "trained" to develop research and alternatives for teaching Geography. They should also "have a sound interdisciplinary education, leadership and democratic management skills, as well as a permanent quest for knowledge, with the ability to construct knowledge that allows them to act as teachers, researchers and participants in the management of educational spaces". This is very much in line with

Study of the Environment presented by Nidia Pontuschka and La Blache explains that:

> The idea of environment, in the various expressions of geography, is used as a correlative and synonym for adaptation. It manifests itself through the series of phenomena that link together and are set in motion by general causes. It is through these causes that we incessantly return to the causes of climate, structure and vital competition, which drive many special activities of forms and beings. (LA BLACHE, 1982 p. 43)

A fundamental part of the interdisciplinary proposal is that it "strengthens interdisciplinary teaching", while it also aims to provide students and teachers with direct contact with a given reality, any environment, rural or urban, that they decide to study (PONTUSCHKA, 2009). Also pointed out by Ângelo Serpa: "The possibility of dividing space is assumed in conjunction with the method and in an interdisciplinary way." (SERPA, 2006)

Há a particular question about the selection of the place to be visited, as well as the formulation of the main questions to be answered in the field research, all the stages of its realization, planning, execution and evaluation, are guided, on the one hand, by "dialogicity" and, on the other, by the awakening of the "curiosity" of all members of the school community (FREIRE, 2000 apud PONTUSCHKA, Nidia N. 2009. P 3).

3.3. Chapter considerations

As is to be expected in the UNIVERSITY - SCHOOL relationship, the participation of the students, directly or indirectly, in the various stages of the Field Work cannot be neglected, because it is the moment when they are close to a complete and curious experience.

At university, the possibility for training in this period is suggested precisely by the visualization of theory in practice, supported by the systematization of the teacher in Field Work through the objectives of research and responsibility in the teaching-learning relationship and with the object of his discipline.

Field records become a real search, presence and participation with the (disciplinary) objects of research and shared theory in dialogue with new knowledge. For this reason, it might be an interesting idea to organize a course on Fieldwork in Geography at UNIFAL-MG, or even to include fieldwork experiences in the course's inaugural classes or in courses on research methods in geography or research methodology.

It is only valid to emphasize that the skills and competencies foreseen (Political Pedagogical Plan Geography degree - UNIFAL-MG) in the formation of the professional geographer (while they are achieved by the student), are perceived more easily by the teachers as involvement / development in the understanding of knowledge through Field Work, achieving in general the expected results.

However, we need to go further. In line with the specific objectives of my research, I propose to discuss a better understanding of the concept of expedition/exhibition, which until then had also been worked on by the authors of the Boletim Paulista de Geografia, but which is best organized in the words of Yves Lacoste:

> The expedition/exhibition seems to be a very effective method of training students for research. (...) The students who have taken part in these expeditions/exhibitions have mostly experienced a certain enthusiasm (especially if they are activists) at the idea of reporting the results of their research to the public. (LACOSTE, 2006)

Therefore, fieldwork is indispensable, but not sufficient, because researching places means, above all, that the researcher is responsible for the result, and communication is an indispensable part of this concept of exposing the results to the community and continuing the research:

> It would be puerile to imagine that the population that is the subject of the research is socially and politically homogeneous: the documents presented in the exhibition are not to everyone's taste, and even those who take a sympathetic view of the students' project are a little embarrassed to see foreigners exposing the life and daily life of this community in diagrams, plans and graphs. This embarrassment at being exposed in this way, even when nothing scandalous is revealed, manifests itself particularly in front of the panel where the image and hypothesis that the group would have in a decade or two is sketched out, in the case where the negative and imagined scenario would be realized. (LACOSTE, 2006. P. 92)

. Fieldwork generally produces a diversity of experiences that are almost never appreciated. To exhibit the results is also to exhibit photos, poems, sketches, and a whole production that begins in that place and about that place:

> Some teachers can *already* draw on these experiences. Firstly, that after a dozen days of intensive work, geography students, even those who are largely beginners, can present a coherent set of information that is largely unprecedented. They have contributed to the awareness on the part of the population studied (but not just the notables and intellectuals)

of a number of problems which, until now, they had not been asking themselves, and whose solutions will determine their future. It is certainly only an initiation into research, but it is the beginning of real research. (LACOSTE, 2006)

This is the aim of a proposed PROJECT called **"POSTAL GEOGRAPHY"**, an initiation to research, fieldwork displayed through a website (expansion of the educational space: through the combination of field experiences and the media), which can reflect on a local-regional-global scale in a more liquid and universal way, (there are still doubts about this "universality"), and be the beginning of real research. There are some specific objectives that must be met when applying the project in a teaching context:

Sharing the different perspectives present in Field Work through the different views of the social subjects involved in the project; Collecting specific data and information about the place, its inhabitants and the relationships they maintain with other spaces; Emersion of disciplinary and interdisciplinary curricular content to be included in the program; Creation of didactic resources based on the records; Dissemination of processes and results. (LOPES, PONTUSCKA, 2009)

The "Postcard Geography" project is also understood as a practice that assumes the production of space in a *self-reflexive* way, recognizing that cultural production and the production of space cannot be separated and that cultural and intellectual production is a spatial practice (MARQUES, 2009):

This is a movement towards the exteriorization of subjectivity, understanding it not as an inner trait constructed subcutaneously, but rather subjectivity as a spatialization, a spatial practice. Instead of associating notions of introspection and isolation with subjectivity, the practice of displacement and transit is incorporated into it (MARQUES, 2009).

POSTAL GEOGRAPHY" AT UNIVERSITY

The Geography - UNIFAL/MG website is the means by which we can view everything from the history of the course to news, useful links and current events. Through this site, Geography displays photos and the places where fieldwork is carried out. Through this portal we intend to create a new one: "Geography Postcards", which also aims to create a collection of data on Fieldwork in Geography, sharing the different perspectives present in Fieldwork through the different views of the students and teachers involved.

"Postcard Geography" should be understood as a means of displaying data and information specific to the places visited (which are real postcards), their inhabitants (residents) and the relationships they have with other spaces. On the website, using an app (examples 1 and 2), the student will translate their map, drawings, sketches and photographs from the Fieldwork into a *postcard,* which will be "engraved" on the app's map. This is an old practice of exchanging postcards, but this time they will be exchanged with other people around the world via the website, which will form a significant "cultural exchange" based on the "exchange of postcards".

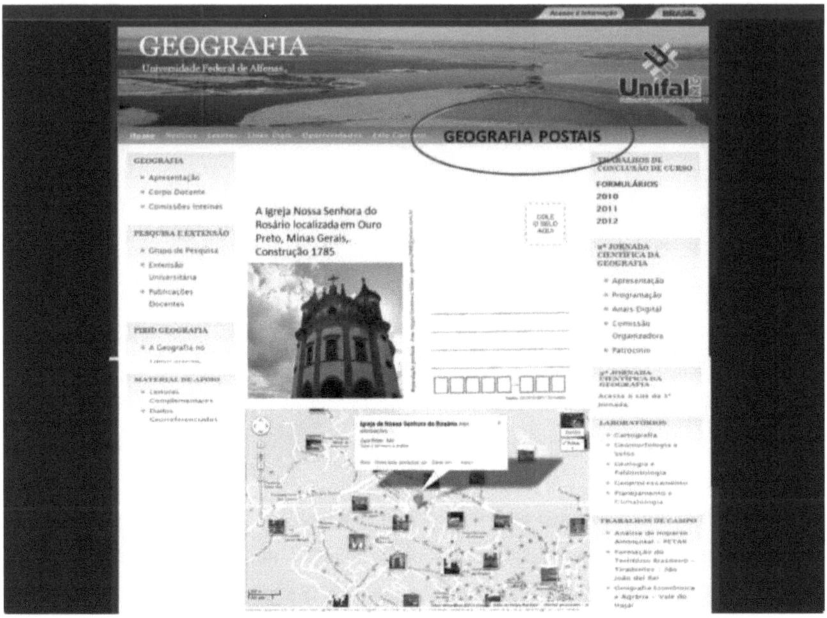

Figure 2 - Example (template) Geography Postcards on the site.

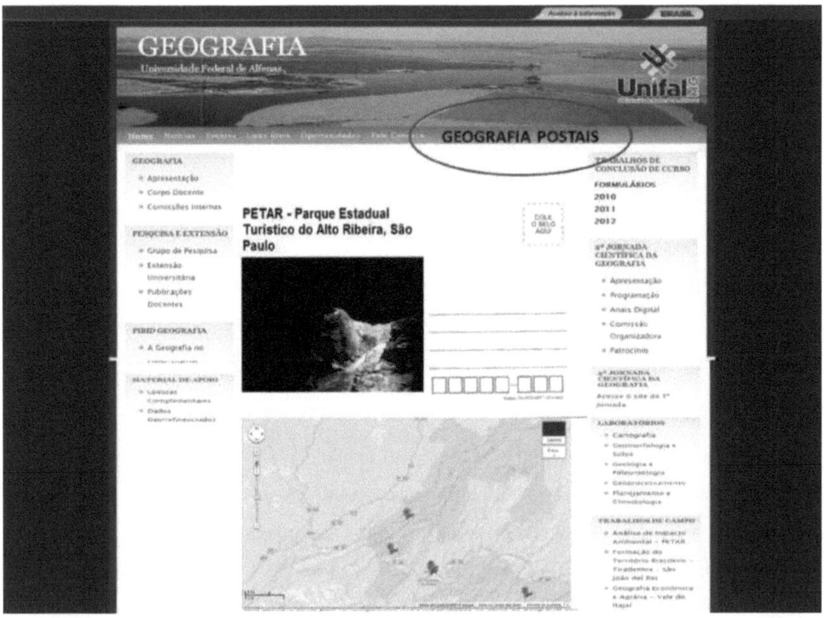

Figure 3 - Example (template) Geography Postcards on the geography website (UNIFAL-MG)

This movement that the Postcard Geography Project will produce is interesting material for Cultural

Geography, with the *website* serving as a passport for exchanging field experiences. It also emphasizes that the university is not the only place where these exchanges can be "held".

In order to strengthen this project, the Computer Science course will be able to take part in the implementation of this *website.* It's about thinking about "portable geography", emphasizing interdisciplinarity and strengthening the junction of areas of knowledge that are located on the same UNIFAL Campus in Alfenas (Campus II - Santa Clara neighborhood). To think of "Postal Geography" is to superimpose "subject and landscape, experience and knowledge, space and spatial practice" on the same site:

The portable condition unleashes the autonomy of action and the freedom to create new vital fictions, practicing a micropolitics - "[...] an analysis of the formations of desire in the social field [...]" - of understanding and proposing forms of coexistence. Portable is something that can be easily transported. And if something can be transported, it can also be shared, distributed and made available. For this reason, portable geographies are a kind of cultural and intellectual production that falls into the category of experimental geography, as defined by Trevor Paglen: geographers, rather than simply studying geography, create geographies that are, in turn, shareable geographies. Portable geographies include a production of space that critically reflects on the mechanisms of the production of space itself. (MARQUES, 2009)

POSTAL GEOGRAPHY IN THE COMMUNITY AND FOR THE SCHOOL

One of the points to highlight in the project is observing the demands of the community where the expansion of the educational space of Field Work in Geography (UNIFAL-MG) should be carried out, which is a premise of the expedition/exhibition concept, considering the problems encountered and proposing solutions that at least change the students' outlook.

Therefore, I still think that the Postal Geography Project could move into a social, extension and research field - identifying the demands of schools and society. In the case of Alfenas, I noticed that the House of Culture and Events is standing idle. No exhibitions. It would be an interesting place to initially house some of our reflections on postal geography.

The place chosen to take them is the Alfenas House of Culture, but it could also be the University itself, thus making the UNIVERSITY - COMMUNITY - SCHOOL relationship. From the educational space in the House of Culture, we intend to invite schools to take part in this "exhibition": **"POSTAIS GEOGRAPHY"**. Field practices will be exhibited through texts, photos, maps, drawings, sketches, in other words, branching out field experiences by exhibiting the results in other environments.

Postcard Geography" in the community and for the school should be understood as a way of displaying specific data and information about the places visited during the fieldwork. The visitors (school and society) to the House of Culture will maintain the purpose of the project, which is to exchange the relationships that the subject maintains with other spaces through the exchange of *postcards*. At the House of Culture, using a *datashow*, students will be able to choose an image from the Field Work and take a photo in the form of a *postcard,* which will be sent to the destination they requested. This is an old practice of exchanging postcards, but this time it will be exchanged with other people who will form a significant "cultural exchange" from the "exchange of postcards", expanding the space for research.

This project will also produce interesting materials for Cultural-Humanistic Geography, with each of those involved working as a "passport" for exchanging experiences, life and science.

To inspire them, the ideas contained in the Geography Postcards Project serve as a starting point for school teachers to think about and, by gathering this material, they will be able to find theoretical material that will help them to build this type of methodology with their students too.

CHAPTER 4

Fieldwork in Cultural - Humanistic Geography

"From where I am, I have already left"

Manoel de Barros

In order to refer to the previous proposal, in this chapter we will look at Fieldwork in Cultural Geography, proposing to discuss mainly the humanistic-phenomenological method in Fieldwork. This move is important because we are going to study *a practice that is associated with a worldview* (BPG, 2006). Is this a defining milestone in Geography Fieldwork and for the answers set out in the "Postal Geography Project"?

The historical line of Cultural Geography presents itself as such based on the diversity of ways of life and landscapes. These contrasts, according to Paul Claval, originate from the disparity of "hunger vs. wealth" in some countries and follow an organization. First of all, it would be explained by nature (1) which imposes on peoples the way they eat, dress and build their homes ("making them peaceful sedentaries or aggressive nomads"). Another point to consider is race (2), which is equivalent to biological differences of a physical nature (strength or aptitude for work) and, above all, intellectual differences, which cause major cultural differences. History (3), another point made by the author, is commented on as if it were a convenient justification for colonization. The diversity of culture and the situations in which it is placed are then related to the "delays" in the evolution of groups. The superior reality (4) is explained as impositions that condition the groups. (5) The unification and disintegration of cultures, apparently condemned to decline by the standardization of technology, rediscovers its dynamism by linking itself to the representations and feelings of identity that are linked to it. (CLAVAL, 2007, p. 10/11/12) In this initial analysis, the author observes the difference in a group's arsenal of tools and the techniques they use to protect themselves from the excesses of the climate and exploit natural resources.

The beginnings of Cultural Geography were German. (CLAVAL, 2007, p. 20) Frederich Ratzel introduced the concept of culture as a key factor in human geography. In the middle of these same years, Ratzel developed a new conception of geography. He observed the lessons of the great German masters of the discipline, Alexander de Humbolt and Carl Ritter, and drew from his naturalist training the idea that the distribution of men and civilizations deserved special attention: he proposed the name of anthropogeography (18811891) to describe the new chapter: 1) anthropogeography describes the areas where men live, and maps them; 2) it seeks to establish the geographical causes of the distribution of men on the earth's surface; 3) it sets out to define the influence of nature on the bodies and spirits of men (BUTTMAN, 1997, p 63).

Ratzel dedicated the 1880s to the study of the cultural foundations of the Earth's regional differentiation, in three works in which Ratzel recognized in peoples an attribute that belongs to their essence, mobility. They master the techniques that ensure their adaptation to their surroundings and depend on their history and level of development (living space). The geography conceived by Ratzel gives an important place to the facts of culture. But this culture is mainly analyzed from a material point of view, as a set of artifacts used by men in their relationship with space. The ideas that underlie it and the language that expresses it are no longer evoked. (CLAVAL, 2007)

Many German geographers were looking for a definition that would offer the advantage of circumscribing a clear object and explaining it, avoiding border disputes with neighboring disciplines. Otto Schlüter (1872-1959) established landscape as an object of human geography, but it is worth defining them:

> An early distinction between the German school suggests that for Ratzel, the geographical study of culture was confused with that of the artifacts used by men to dominate space; for Shlüter and the majority of German geographers from the first decades of the 20th century, it is the mark that men impose on the landscape that constitutes the object of geography: to apprehend this organization, to describe what has since been described as the morphology of the cultural landscape and to understand its genesis. (CLAVAL, 2007, p. 24)

American cultural geography (Carl Ortwin Sauer - 1889 - 1975) worked on top of German logic, which until then had been interested in landscape studies and the relationship between culture and space. In the United States, the dominant school from 1910 until the Second World War totally ignored them. Very concerned with rigor, it paid a lot of attention to data collection and representations, generally confusing cultural geography with human geography. For him, however, geography is limited to what is legible on the surface of the Earth and, like the German geographers, ignores the social and psychological dimensions of culture (CLAVAL, 2007), which defines geography:

> Our naively selected section of reality, the landscape, undergoes multiple changes. This contact between man and his changing home, as expressed through the cultural landscape, is our field of study. We are concerned with the importance that the place has for man, and also with the transformations it imposes on the place. In short, we deal with the interrelationships of the group, or cultures, with the site, as expressed through the various landscapes of the Earth. (Sauer, 1974, 1 ed. 1925. "The morphology of landscape" University of California Publications in Geography apud CLAVAL, 2007)

Like his contemporaries, Sauer sees culture primarily as the set of tools and artifacts that allow man to act on the outside world, but he goes further than them: culture is also made up of associations of plants and animals that societies have learned to use to modify the natural environment and make it more productive, and this is one of the major traits that cultures should be judged by. For this reason, Claval warns that in order to make use of Sauer's theories, the geographer must have a solid naturalist background (CLAVAL, 2007, p.31).

The cultural dimension of French geography emphasizes the *genre of life* and *landscapes*, based on the legacies of Alexandre de Humbolt, Carl Ritter and Friedrich Ratzel: this is the case of Paul Vidal de La Blache. La Blache starts from the concept of human geography proposed by Ratzel (study of the influences of the environment on human societies). In order to analyse it, he is also interested in the set of techniques and tools that men make to transform the context in which they live and that these elements do not make sense if they are not understood as components of the genre of life, this being a specificity of geography: the studies of places, but the analysis of genres of life shows how the elaboration of landscapes reflects the social organization of work, gaining in La Blache's version a social and ideological dimension that are linked to its ecological aspect.

The notion of genre of life thus introduces a logic into French human geography that encourages the integration of increasingly varied and complex behavioral aspects into its field. Naturalistic in its origin and its justifications, it quickly drifted towards more humanistic positions.

In the first half of the 1970s, we can highlight the names of Tuan and Butttimer as those who most contributed to the search for an identity for humanist geography. These authors were pioneers in the use of the concepts of place and the lived world, both associated with a phenomenological existentialist theoretical basis, which would later allow their work to be identified as humanist (HOLZER, 2003).

Through humanistics, Yi-Fu Tuan*, one of the exponents in the elaboration and systematization of Cultural - Humanistic Geography in the 19th century, points out that "this method reflects on geographical phenomena with the aim of achieving a better understanding of man and his condition. 19th century, points out that "this method reflects on geographical phenomena with the aim of achieving a better understanding of man and his condition, thus seeking an understanding of the human world through studies of social relations with nature, their geographical behaviour as well as their feelings and ideas about space and place" and also formalizes the presentation by criticizing scientific approaches to the study of man, which tend to minimize the role of human awareness and knowledge. Humanistic Geography, in contrast, specifically tries to understand how geographical activities and phenomena reveal the quality of human awareness." (TUAN, 1976, p. 143)

The themes for the scientific analysis of Humanistic Geography are available in the book (pages 147-155), so as to offer a new way of looking at geographical phenomena, Tuan presents them as themes of general interest to geographers:

1. **Geographical knowledge** is an "animal instinct", necessary for survival. In contrast, the geographic knowledge cultivated in academic departments is highly knowledgeable and specialized. We know little about how "spatial skills" develop from group to group, and the differentiation of

more primitive and more evolved cartographic skills, for example, from another more specialized group with a lower degree of apprehension of geographic knowledge;

2. **Territory and Place:** Territory is an idea of an area communicated by paths and places. "Human places vary greatly in size and lead us to reflect on the emotions and thoughts attached to place. (...) How a mere space becomes an intensely human place is a task for the humanistic geographer, and to this end he appeals to distinctly humanistic interests, such as the nature of experience, the quality of the emotional attachment to physical objects, the functions of concepts and symbols in creating the identity of place."

3. **Human agglomeration and privacy:** The impact of high population density on quality of life has attracted the attention of scientists and social planners. Can agglomerations produce tensions, leading to illness and anti-social behavior? According to humanistic theory, culture mediates between density and behavior, but the degree to which this occurs is a major challenge. The distinctive humanistic approach is to describe the qualities of emotion experienced in specific cases;

4. **Way of life and economy:** All human activities seem to be economic and functional, in the sense that they support the social system, without which people could not live. The humanistic perspective on economic life can be presented as an answer to two questions. Firstly, it answers the meaning of the terms "life-sustaining activity" and "way of life", suggesting a moment of reflection on the need to adapt to the environment. This process affects life's behaviors with different intensities. In the modern world, the way in which man's life behavior is identified as economic is clear, especially when observing his production and exchange of material goods, and varies widely from society to society and within each society.

5. **Religion:** Religion is also present to varying degrees in all cultures. A humanistic geographer seeks to consider that a religious person is one who seeks coherence and meaning in their world, and a religious culture is one that has a clearly structured worldview. A humanistic approach to religion would require us to be aware of the differences in human desire for coherence, and to note how these differences are manifested in the organization of space and time and in attitudes towards nature.

Still in dialogue with Tuan's (1976) theory, the themes are suggested for a local dimension that consolidates the ramification in regional analysis and is also based on the understanding of history for Humanistic Geography:

> History is not just the passage of events, but their conscious reconstruction in the group's memory for current needs. However defined, history plays an essential role in the human sense of territoriality and place. The humanist must know the factual past of the place, but understand that the past does not determine the current identity of the place. Its current

identity is being created (...) through the selective use of its past. (TUAN, 1976, p 155)

Thus, "lived description" may have been one of the great achievements of regional geography. Understanding the definitions of Humanistic Geography, he places regional geography as a study of the type: "study of art", because according to the author, humanistic perception succeeds in capturing the essence of place (the region is the result of one person's art multiplied several times). (TUAN, 1976, 156).

In search of the art of doing geography, the humanist geographer, according to Tuan, begins with linguistic skills (and not just in the sense of writing well), within which he makes use of the ability to recognize the nuances of language and the ambiguous meaning of key words. You need to develop a sensitivity to language (...) "read between the lines and hear what hasn't been said." Or again, according to Claval, culture is the inheritance and result of the game of communication, and to a large extent it is made up of words, articulated in discourse and realized in representation. But it is still worth highlighting the mode of arrival and the approach that, according to Valéria de Marcos:

(...) the main source of data is the testimony of the people involved and the only way to guarantee that this testimony is true is to gain the trust of the research subjects. Unfortunately, a lot of research is done without considering these issues. It is still a practice among many researchers to arrive at the study area with endless questionnaires constructed in the office, "machine-gun" the "objects of study" with their often incomprehensible questions and walk away. Often the community never hears from the researcher or the study he has carried out. Now, faced with the repetition of these situations, the community learns to no longer trust strangers and, little by little, builds up its own defenses: sometimes it closes itself off to future researchers and prevents new research from being carried out, sometimes it prepares a "ready-made" discourse, "packaged" like the questionnaires applied, to be "answered" by the next researcher who arrives. (MARCOS, 2006. p. 114)

Geography from a humanistic perspective makes use of phenomenology, converging field research to clarify the meaning of concepts, symbols and aspirations as they relate to space and place, which in turn rests on the interpretation of human experience in its ambiguity, ambivalence and complexity (TUAN, 1976, p 162). Oral history is an important technique:

This theoretical reflection would become more sophisticated over the following decades as "Cultural Geography", as an academic product of this reflection, became the discipline most taught in North American geography courses and the one that generated the most field research. In this context, the theoretical contribution was no longer discussed, but the fundamentals of observing the factor in its relations (lived space) remained the same (HOLZER, 2003).

Tuan stated that "space" and "place" are the concepts that define the nature of geography. He also introduced time as a concept in constant interaction with space. The author defended the importance of his research: "All academic work increases the field of consciousness. Humanistic studies also contribute to self-awareness, to man's growing knowledge of the sources of his knowledge. In every major discipline there is a humanist subfield which is the philosophy and history of that discipline.

Through this subfield, for example, geography or physics get to know themselves, that is, the origins of their concepts, assumptions and biases in the experiences of their pioneering sages and scientists." (TUAN, 1979 apud HOLZER, 2003)

Space from a humanist perspective will refer to spatial feelings and experience. Experience is defined as: "the totality of means by which we come to understand the world: we come to know the world through sensation (feeling), perception and conception." (TUAN, 1979 apud HOLZER, 2003). For Tuan, "The importance of 'place' for cultural and humanistic geography should be obvious:

> Like functional nodes in space, places are subject to the techniques of spatial amphilysis. But as a single, complex whole - rooted in the past and growing into the future - and as a symbol, place cries out for humanist understanding. In the humanist tradition, places have been studied from historical and literary-artistic perspectives (HOLZER, 2003).

Contemporary work in cultural geography has led us to reflect on the various strategies of resistance developed by subordinate groups to challenge the hegemony of those in power. Rejecting what they label the elitist and outdated preferences of traditional cultural studies, they explore a wide range of subcultures. For a final reflection:

> As a critical cultural practice, geography is dispersed, allowing us to think of other ways of knowing space through a set of *artisanal practices, a* term proposed by Boaventura de Sousa Santos. The imaginary of craftsmanship incorporates the portability of knowledge, the small scale of production and the obligatory intersection with a personalized point of view and way of doing things. Craftsmanship is thus able to bring together science and art interested in the study of space and discuss its mechanisms of expression and dissemination. Boaventura de Sousa Santos uses the term artisanship of practices to designate knowledge that can be detected through the sociology of absences and the sociology of emergencies, in other words: knowledge that is not part of the hegemonic knowledge exuded by the world's centers of power - exclusionary scientific power, the power of economic domination, the power of cultural colonization. The artisanship of practices is located in non-epistemological fields in the conventional sense of the term, that is, in places other than those such as universities and scientific research centers. (MARQUES, 2009)

4.1 Chapter considerations

Based on these humanistic and phenomenological reflections, fieldwork makes sense because it is a process of identification mediated by cultural and social signs, insofar as it brings the scientific value of field research to the place, and in humanistics it is an absolute method of exploration.

In Brazil, we should emphasize that the richness of cultural diversity and its participation in the production of contemporary space are the subject of several publications. To this end, in addition to pointing out cultural diversity, it is necessary to value retained cultural traits. These are important points that Brazilian geographers such as Roberto Lobato Corrêa, Zeny Rosendahl, Welther Holzer and other names in Brazilian Cultural Academic Geography have dedicated themselves to investigating.

Conclusions

As is to be expected in this type of research, I spend a large part of the text reflecting on the theoretical thinking, trying to highlight the methodology envisaged, the method of analysis and, in a way, contribute practical alternatives to the experiences in Geography Fieldwork.

The initial perception of the importance of Fieldwork is thus understood as a fundamental practice in the formation of geographic science, partly due to the observations and field reports - evocation of the landscape and man in the different meanings of adaptation to the environment - and partly juxtaposed by the concern to suspend the (im)perfect dialectic between man and nature and explore, in true reality, geographic knowledge, attributing qualities to it.

However, fieldwork alone will not "save" the discussions specific to geography. Especially at university level, even when it comes to the "specific" in Geography, if we want to achieve the designation of geographic science, we need to practice responsibly with the objects of study and, above all, value the experiences of the subjects in Fieldwork.

We tend to believe in valuing students, who often don't feel prepared for professional work (as they are still in training), a fact that ends up preventing the full development of Field Work. I believe that this should be an activity organized from a curricular design based on "working" in teams, contributing to the application of knowledge in initially collective research, highlighting the exposition of the groups and the synthesis of the groups, as well as the exposition of the consequent conclusions regarding the Field Work, as the beginning of a real research!

Questions of teacher training, however, involve "overcoming" the subject's own trajectory because he or she carries with him or her "baggage", a spontaneous conception. In this way, we should try to work on the individual's potential to overcome limits, between myths, habits and try to invest in creativity and wisdom in order to critically organize the various fields of activity of the geographer in an increasingly dynamic world...

It is clear that there must be preparation not only to relate theory and practice, but above all there are many factors that influence good practice both at university and at school, which means that the value of scientific rigor (theory and practice), whether you like it or not, is supported not only by theoretical input, but also by responsible field experience.

The importance of the systematization of the humanistic method in Field Work has been shown, both at the university and at school, to be a practice that proves its own identity - the idea of the heritage of identity. For this reason, the importance of Fieldwork arises in an environment of new communications and technologies that can empower teachers to think locally and act globally. Experiences, made shareable through exhibition, project meaning onto reality, the circumstances of which are presented in life as it is in everyday life.

So, thinking about the prepositions of the Postcard Geography Project means, above all, bringing together these field reports so that they don't get lost. It was the way I invested in promoting a meaningful experience (my own experiences) with direct contact with knowledge, constituted through knowledge and experiences that intertwine. This methodology of geography is the potential in the process of experience and cannot be "retained" only by university students; extension is essential when we talk about the need to build something in Alfenas based on our field experiences.

Fieldwork and the Postcard Geographies Project therefore become an invitation not to accept professional work that is submissive to institutionalization, be it private or public. It's disturbing to think of a training course in which concepts seem to be "buried" in a limited vision, and which, thinking of my own professional "role", advocates "wearing the uniform of the institution", in the sense that it prevents innovation and the development of new actions.

Finally, I would like to emphasize that the methodology used in this research, through the Boletim Paulista de Geografia and all the "paths" (I would like to take advantage of the guidance of Professor Flamarion Dutra Alves, who supported me at times when I was unsure, as well as the professors responsible for the fieldwork, who had an incredible experience in my training) assigned to the study of fieldwork, are the primary sources for the analysis of the proposed theme. Although I analyze them as such, I understand that it is not enough, due to the complexity that each Field Work is organized within Geography, generating diverse experiences that reflect on the "knowledge-building" potential that this type of activity assumes.

As a geography student and therefore a future teacher, I have dedicated myself to the (re)construction of geographic thinking as a possibility of understanding the "art of doing geography" through the importance of Fieldwork. I point out its relevance, contemporaneity and effectiveness, understanding it as: deepening and theoretical construction in geography and the exhibition, as the new look at reality, attentive to the naivety of the students and convinced that the expedition is only just beginning...

REFERENCES

ALVES, Flamarion Dutra. Theoretical-methodological trajectory of Brazilian agrarian geography: production in scientific journals from 1939 to 2009. Thesis (Doctorate) Institute of Geosciences and Exact Sciences, Paulista State University - Rio Claro Campus. 2010.

BPG. Boletim Paulista de Geografia / Sao Paulo Section - Association of Brazilian Geographers. - n° 1 (1949) - Sao Paulo: AGB, 138 p, 2006.

SERPA, A. S. P. Fieldwork in Geography: A Theoretical and Methodological Approach. In: Boletim Paulista de Geografia, v. 84, p. 7-24, 2006.

BAITZ, Ricardo. The implication: a new sediment to explore in Geography? In: Boletim Paulista de Geografia, v. 84, p. 25-50, 2006.

ALENTEJANO, Paulo Roberto Raposo; Otàvio M. ROCHA-LEÂO. Fieldwork: an essential tool for geographers or a trivialized instrument? In: Boletim Paulista de Geografia, v. 84, p. 51-68, 2006.

VENTURI, Luis Antônio. The role of technique in the process of scientific production. In: Boletim Paulista de Geografia, v. 84, p. 69-76, 2006.

LACOSTE, Yves. Research and fieldwork: a political problem for researchers, students and citizens. Revista Heródote n° 8, Oct./Dec. pages 3 to 20, 1977. In: Boletim Paulista de Geografia, v. 84, p. 77-92, 2006.

KAISER, Bernard. The implication: a new sediment to explore in Geography? In: Boletim Paulista de Geografia, v. 84, p. 25-50, 2006.

MARCOS, V. Fieldwork in Geography: reflections on an experience of participant research. In: Boletim Paulista de Geografia, v. 1 n 84, p. 105-136, 2006.

TUAN, Yi-Fu. Humanistic Geography. p.143-164. In: CHRISTOFOLETTI, Antônio. Perspectives on Geography. Sao Paulo: DIFEL - Difusao Editorial S. A, 1982.

BUTTIMER, Anne. Apprehending the dynamism of the lived world. p. 165-193. In: CHRISTOFOLETTI, Antônio. Perspectives on Geography. Sao Paulo: Difel, 1982.

CLAVAL, Paul. Introduction; Genesis and evolution of cultural interpretations in geography. p.9-40. In: CLAVAL, Paul. Cultural Geography. 3.ed. Florianópolis: Ed. da UFSC, 2007. Available: https://sites.google.com/site/flamariongeografia/geografia- cultural

HOLZER, Werther. The concept of place in cultural-humanist geography: a contribution to contemporary geography. p.113-123. In: Geographia. ano V, n.10, 2003. Available: https://sites.google.com/site/flamariongeografia/geografia-cultural

LA BLACHE, Paul Vidal de. The characteristics of geography. p.37-47. In: CHRISTOFOLETTI, Antônio. Perspectives on Geography. Sao Paulo: Difel, 1982. https://sites.google.com/site/flamariongeografia/historia-do-pensamento-geografico

LOPES, Claudivan S.; PONTUSCHKA, Nidia N. Estudo do meio: teoria e pràtica. Geografia (Londrina) v. 18, n. 2, 2009. Available at:

http://www.uel.br/revistas/uel/index.php/geografia/

MARQUES, Renata Moreira. PORTAL GEOGRAPHIES: art and spatial knowledge. Thesis (Doctorate) Federal University of Minas Gerais. Institute of Geosciences UFMG, 2009.

POLITICAL PEDAGOGICAL PLAN FOR UNDERGRADUATE GEOGRAPHY. Federal University of Alfenas - UNIFAL/MG. Alfenas, 2O11. Available at: < http://www.unifal- mg. edu.br/geografia/?q=node/3 >

SPOSITO, Eliseu Savério. Geography and philosophy: a contribution to the teaching of geographical thought. Sao Paulo: Editora da UNESP, 2004.

APPENDIX I - Summary table (complete)

BPG: concepts and themes present in the articles of the Boletim Paulista de Geografia (NÙMERO 84. SÂO PAULO-SP, JUL.2006 - O Trabalho de Campo)

Summary table

Article	Education/Biography	Geographical School / Method of Analysis	Themes/debates	Concepts	References
1- Ângelo Serpa **(Angelo Szaniecki Perret Serpa)**	Brazilian. He is a full professor of Human Geography at the Federal University of Bahia; PhD in Landscape and Environmental Planning from the Universitaet Für Bodenkultur Wien (1994), with post-doctorate studies in Studies in Outer Space Organization and Urban-Regional and Landscape Planning at the University of Sao Paulo (1995 - 1996) and in Cultural and Urban Geography at the Université Paris IV (Sorbonne/2002-2003) and at the Humboldt Universitat zu Berlin (2009).	Critical Geography and Cultural Geography - phenomenological.	- Theoretical and methodological reflection on the importance of fieldwork in the construction of geographic knowledge in contemporary times; - Fieldwork as a specific discipline of geography; - Articulation of concepts, theories and methodological procedures in geography; - Fieldwork is the possibility of cutting out, analyzing and conceptualizing space, according to the questions, goals and objectives defined by the research subject; - Fieldwork as an instrument for overcoming dichotomies in geographical science.	Space; Significant spatial cross-section e; Landscape; Vertical Landscape; Cultural Landscape; Theory-method; elements; totality	**Milton Santos, Paul Claval, Yves Lacoste, Inà Elias de Castro; Roberto Lobato Corrêa**; Hannah Arendt; Odeiblier Guidugli; Racine, J.B.;Raffestin, C.;Ruffy, V.; Màrcio Mendes Rocha; Edward W. Souza; Marcelo José de Lopes Souza; Eliseu Saverio Sposito, **Henri Lefbvre.**
2-Ricardo Baitz	Geographer FFLCH/ USP. Lawyer from PUC/SP and Postgraduate in Public Law from ESA/SP.	Critical Geography.	- A implication as a methodology for fieldwork; - The relationship between implied subject and implied object; - Approaching the object/researcher as an active subject; - The neutrality of science; - Implications and	Implication subject-object; institutions exhibition	**Henri Lefbvre; Remi Hess; René Lourou**; Edgar Morin; Friedrich Nietzsche; **Paulo Freire**; Piaget
			pedagogy; -Implication and participation;		
3- Alentejan, Paulo R.R	Brazilians. Teachers Department of	Critical Geography; (Dialogue	- Fieldwork and geographical theories; linking theory and	Landscape evolution models; Monitoring of	**Yves Lacoste; Bernard Kaiser, Antônio Tomaz**

& Rocha-Leao, Otâvio M.	Geography UERJ-FFP	between physical geograph y human geographer)	practice; - The importance of fieldwork for geography research and teaching; - Fieldwork as a tool at the service of the geographer; - AGB as a disseminator of science based on fieldwork practices; - Dangers that The trivialization of fieldwork - Theory and method, - Dichotomies in geography, - Transformative action,	field; scale; theoryand practice;	**Junior, Mauriciode Almeida Abreu, David Harvey, Manoel Corrêa de Andrade, RichardJ . Chorley, Antônio Christofoletti, AnaLuiza Coelho Netto, JoaoRua , Armen Mamigonian, Milton Santos, ArthurN . Stralher, Dirce M. Suertegaray, Mao Zedong.**
4- Luis Antônio Bittar Venturi	Brazilian. Professor at USP's Geography Department. He works in the areas of Geography of Natural Resources, Theory, Method and Techniques of Field and Laboratory Research in Geography, topics on which he has given lectures, published articles and books, such as "Geografia - prâticas de campo, laboratòrio e sala" (Geography - field, laboratory and classroom practices).		- Method and technique are complementary; - Technical work vs. scientific work; - Techniques evolve according to man's needs; - technical mastery = complexity of new technologies vs. the option of using instruments simpler. - Laboratory techniques - Fieldwork techniques = planning flexible.	"Thinking and doing", "scientific work", technological instruments- "seduction of technology". laboratory, fieldwork, "cabinet", "object" of study,	**T. R. Giles, Yves Lacoste, Bernard Kaiser, G.G. Granger, A. G Cunha,**
	classroom ' (Ed. sarandi, 2011)				
5- Yves Lacoste	French geographer. At the beginning of 1970, he launched the magazine Hérodote, which for the next thirty years sought to reveal the hidden face of geography, namely its political character. In 1976, Lacoste wrote the book: "Geography - it serves, in the first place, to make war." In 1980, Lacoste and the members of Hérodote devoted	Critical Geography	- Research and fieldwork: problems of political order; -The role of university; - collective research vs. individual research; - the social role of the researcher; - the social role of the researcher; - the social role of the researcher. propose solutions/communicate the results, what will it	Responsibility of the researcher; research results; expedition/exhibition; subject-research object; landscape; evocation of landscape; man-nature relationship;	

	themselves to the problems of internal and external geopolitics, on a regional, national, continental and international scale.		provoke in the group studied? -Commitment with the organization; - Theoretical + practical+ training articulation of scale; - The "power" of research; - The role of the state and big business; - The technology monopoly; - difficulties budgets;		
6- Bernard Kaiser	French geographer, professor at University of Toulouse-le-Mirail, and one of the creators of the Active Geography, a movement that defended geography as an active and changing factor in the world, a movement that influenced the creation of the Critical Geography.	Active Geography	- The importance of field research; - researcher = citizen; -typology of fieldwork for the social survey; -specific geographic field research? - methodology proposal: local bibliographical references + theoretical training + repercussions of the hypothesis; - social proposal: conflicts and problems; class struggles; - everyday life; - justification for research and the responsibility of the researcher; -valuing the use of local people's knowledge;	Subsystem; situational analysis, local analysis, historical analysis; local/globa l "	**MaoTsé Tung, Yves Lacoste,**
7- Valéria de Marcos	Brazilian. Graduated from USP; Professor Doctorate Department of Geosciences CCNE UFPB and PPGG CCEN UFPB.	Geography Humanistic-Cultural.	- Discusses the importance of fieldwork in geography; participatory research as a methodology and experience; - How the research was carried out; - the researcher's commitment to the community; - the attitude of the researcher when approaching the community; - the difference in temporal understanding for the	Participant research; community production a; teaching-learning process, educational action	**Mao Tsé Tung, José de Souza Martins, C. R. Brandão, E. P de Godói, S. Cardoso, J. G. C Magnani, A. U. de Oliveira**

			researcher and the community being studied; the importance of the gaze in carrying out research;		

MIX
Papier aus verantwortungsvollen Quellen
Paper from responsible sources
FSC® C105338

Printed by Books on Demand GmbH, Norderstedt / Germany